The Secret
OF Energy

"*The Secret of Energy* is not an esoteric book but a layman's book on the basic principles and function of energy in our daily lives."
-Dr. Gary L Storkan

The Secret
OF Energy

A Spiritual and Scientific Tool
For Finding Ultimate Happiness

Marcie Martinez, B.S. Ch.E., M.S. M.Sc.

Foreword by Dr. Gary L. Storkan

Printed in the United States of America

First Printing, 2018
ISBN-13: 978-1981306206
ISBN-10: 198130620X

Nature's Presence Publishing
www.naturespresence.net

For copies of this book and future books in *The Secret of Energy* series as well as energetic healing tools visit http://secretofenergy.com.

First Edition

This book is dedicated to all naturopathic doctors, chiropractors, homeopaths, and energy healing practitioners around the globe.
-Marcie Martinez

Contents

Foreword

"The Secret of Energy is not an esoteric book but
a layman's book on the basic principles and
function of energy in our daily lives."
-Dr. Gary L Storkan

In this book, you will learn about the many areas of
your life in which energy plays a significant role, and some
well developed techniques to help you work with the
distortions that can occur on a daily basis.

In my own practice I see, daily, patients with
significant metabolic and physical problems. When tested,
these people often reveal very high levels of stress. I have
found that while in this state of emotional and energetic
disturbance, their bodies are unable to maintain normal
function that manifests as metabolic disease. Once the
"triggers" (as I call them) that are causing this energetic
distortion are removed I have seen the body make
seemingly miraculous recovery. It really boils down to
allowing the body to reorganize its energy and function
more efficiently.

In her first chapter, Marcie outlines the scientific explanation of energy and how it is measured by the scientific community. I think this is important to mention because I believe many people think of "energy" as a "New Age" concept without any substantial foundation. She also mentions how it is connected with our spirit and subconscious mind (however you might relate to that concept).

She then moves into chapters on how energy can be manipulated to affect our health and lives. She describes the importance of sleep, the electromagnetic pollution, and cell phones that have become such an important part of our lives. She states that these necessities can have either a positive or negative influence on our energy systems and consequently the health of our bodies.

In subsequent chapters, Marcie describes many personal experiences with various modalities that have impacted her life and in many ways led her to writing *The Secret of Energy*. I too have experienced some of the modalities she describes and know personally their powerful influence on my life and health.

Throughout the book, she refers many times to the subconscious mind and the powerful influence it has on our lives. She spends much of the remaining chapters describing techniques to influence the effect of the subconscious mind in our lives. She devotes a significant portion to Dr. Uwe Albrecht, a German medical physician who developed such a system called "Innerwise". I am not personally familiar with this system but after reading this book, I plan to investigate it for myself.

In the last few chapters, Marcie talks about the importance of intuition and the importance of how your intuition can guide you through life (helping you with your health and protection). She describes how even our thoughts carry enough energy to have an influence on other people, our community, and even as physical as water. This is described by the Japanese scientist, Masaru Emoto in his book *Hidden Messages in Water.* In this book, Dr. Emoto illustrates the effects our thoughts can have on the quality of water. Marcie also describes that even though energy is energy, it is the frequency of that energy that can have the positive or negative effect in our lives.

Marcie finishes the book with a chapter on the Schumann Resonance. In the 1950s, a physicist named Otto Schumann discovered that the earth has its own frequency. This has now become know as the Schumann Resonance. It has since been used by many scientists to track the Earth's "heartbeat", so to speak, and its effect on time.

I believe *The Secret of Energy* is an important book to read for anyone who may not have a deep understanding of the meaning and effect energy plays in their life. It will give you a very good and basic understanding of how energy affects one's health, emotions, interpersonal relationships and universally all aspects of your life. Have fun experiencing how you can consciously affect this energy both in your subconscious and external environment.

Gary L. Storkan, DC
Active practitioner for 33 yrs

Acknowledgments

First and foremost I must acknowledge my Source of being - God. For without access to the Holy Spirit I would have nothing and I would be nothing. It is from this Source that all knowledge comes and from this well of energy that our intuition speaks. The spark of life, which powers all bodily functions, thoughts, words, actions and beyond is only possible because of the gift of the soul.

I am immensely grateful to my husband Gil for allowing me the time and space to work on my passion of researching and writing. Without his support I would not be able to follow my dream of helping others through natural health and helping myself through the therapeutic act of writing.

The information in this book was obtained from many sources over the course of years; some of those sources are experts in the topics discussed herein, namely energy as a scientific concept, others are holistic practitioners, and even more are spiritual leaders and lay people alike who are deeply connected with themselves on a spiritual level. I have learned the importance of energy in many situations from many people whether or not they knew they were teaching me about these things.

I am grateful for my education in engineering and science as well as my experience working with radioactive materials and in other scientific roles at the Los Alamos National Laboratory, for without this background I might not be as open to the concept of energy as it functions within our bodies and in our lives. Lessons can be found everywhere, even in the places we feel we do not belong. Understanding this is a lesson in itself allowing us to remember to always be present wherever we are.

I would like to thank Linda Villaseñor for introducing me to Dr. Uwe Albrecht and Innerwise. Dr. Albrecht's insight and knowledge of energy, energy healing, the subconscious mind, and intuition along with his method of teaching these concepts have proven invaluable to me in my own personal life. His workshop not only taught me the importance of intuition but it also opened my eyes to the power within and the incredible possibilities for energetic healing.

To the many holistic practitioners that have treated me for various ailments using solely energetic concepts, I thank you from the bottom of my heart for showing and sharing with me a better way of healing. My body, mind, and spirit appreciate you.

Drs. Scher, Dabby, Itomar, and Indy from The Scher Center for Wellbeing require special mention here, as they have not only opened up my spine to accepting all things, they have also shown me that transformation can happen from the most unlikely sources and heartache can foster incredible growth and opportunity.

To Dr. Dee Blanco, thank you for taking such wonderful care of my little dog with your expertise in homeopathy and thank you for the conversations and for your love of all fur babies. If only there were more like you and if only more pet parents knew about you there would be much less suffering in the domesticated animal kingdom.

To my Dr. Storkan, thank you, thank you, thank you. I am forever grateful for your review of my book and for providing the Foreword. Indeed I feel like I am standing on the shoulders of giants, and you are one of them. This is but a fraction of the gratitude I have for all you have done for me in helping me to appreciate my body's innate wisdom and for being the conduit toward healing of my most difficult ailment - my allergies. As a child I spent many weeks and even months out of each year with my eyes red, itchy and watery, and my nose stuffed. This horrific fate followed me into adulthood until I was doomed to daily pharmaceuticals for relief. That is until I found you. You once said I was your most difficult case yet here I am, breathing freely with only minimal sensitivity to the things that once plagued me to the point of having no energy to do the things I loved and you did this by clearing energy blockages. Your ability to foster healing in your patients is bar none.

Finally, to the people who have read my work and given me feedback, thank you so much for your support and encouragement. I hope this book does not disappoint and, more than that, I hope it helps you on the journey through the twists, turns, ups, downs, straightaways and mazes of this thing we call life.

Introduction

When it comes to the area of natural health and healing it is easy to get wound up in simply looking at the body's ability to heal itself and studying the natural compounds which help speed the process of self-healing. What most people do not realize is there is a secret not only to natural health but also to life itself and, unfortunately, if you do not understand this secret or, especially, even know it exists then you are at a serious disadvantage.

As inescapable as it is that the body is capable of incredible things, not the least of which is its ability to heal itself, many people still believe in the rhetoric coming from the mainstream regarding the importance of "medicine" when it comes to overcoming illness. In actuality we have all been duped. The dumbing down in all aspects, but especially with regards to our most important asset – the body – was gradual, incredible, and brilliant.

When you hear something over and over again you tend to believe it subconsciously regardless of whether it is true or not or even if it doesn't make any sense at all. In all likelihood this is why we have accepted, wholesale, the notion that we need doctors, hospitals, and drugs in order to be healthy. The nearly comical irony of this is the very medical system on which we rely for health is the same system responsible for many illnesses and death. In addition, the very medical system into which we put so much faith uses the phenomenon of energy against us more than we realize.

This, of course, is not to say that doctors in general are deliberately trying to sabotage our health. While there are good and bad people in all walks of life including the medical field it is obvious most doctors choose their profession ultimately to help others and even save lives. Therefore to suggest they even realize they are essentially trained to keep everyone sick would be unfair and generally false. It would be safe, however, to assume doctors are generally as duped as the rest of us when it comes to the field of so-called health care.

There have been some incredible advances in medicine thanks to conventional medical doctors and researchers and therefore this is not intended to disparage the allopathic system altogether. In fact, mainstream medicine in emergency situations and even in situations where a patient is dealing with, for example, a very aggressive cancer, has proven to be invaluable and even critical to many people. Moreover there are many mainstream medical doctors who are now embracing "alternative" treatment modalities either completely or as an integrative program for treating patients.

Still, there remains an air of arrogance within the mainstream medical system, which is a great danger to society as a whole. This appears to be conventional medicine's Achilles heel, as even among laypeople and professionals alike doctors are placed on a pedestal and even treated like celebrities by some. Given this, one thing I find ironic is how many mainstream medical doctors shun the idea of energetic healing as nonsensical when, in fact, it is highly scientific and could provide answers to many mysteries of the human body. Fortunately there are some doctors who are branching into the energetic realm with great success. The tide is turning and it seems that soon enough the mainstream will catch on to what doctors in the East have known for millennia.

Unfortunately the speed at which allopathic medicine is catching up to the innate wisdom of the body seems to be more of a snail's pace. Although many doctors are still prescribing toxic treatments what is most interesting about the role of doctors in western health care is not their active undermining of the body's natural abilities but the subconscious way they deal with patients, which likely has more effect on a person's health than even what they recommend to be put into their bodies. What I am talking about is a type of subliminal message doctors regularly send to their patients' subconscious minds likely without even realizing they are doing it themselves. In essence it is one person's subconscious speaking to the other person's subconscious.

One of the major problems with mainstream medicine is the promotion of the victim mentality, which divests patients of responsibility for their ailments or discomfort. Instead of empowering patients with the knowledge that they have more to do with their state of health than they realize doctors regularly fail to place responsibility on their patients to make necessary lifestyle and even mindset changes. As a result people continue down the same dangerous path of self-harm with the idea that, should they fall ill, they can just get a prescription and be on their merry way.

Although most people still blindly follow their doctors' recommendations there are many others who are beginning to question the recommendations as they are finding them to be ineffective and, at times, even more harmful than if they had done nothing at all. Moreover people are beginning to trust their own common sense more than ever. They are beginning to open their eyes to many truths not just in the medical field but in other areas as well.

Many people sense a change in the air, a convergence of many things so to speak. There are many people who believe the "end times" are near. Whether you believe in religion or not I think it is safe to say times are changing. It is my belief the "end of the world" is simply the end of life, as we know it. If you consider the name of the final book of the bible, the book of the apocalypse or "revelation" you know there is to be a great revealing of truth. I am seeing it in all aspects of life, including in the field of medicine.

It seems almost weekly I am made aware of some "new" idea that, ironically, is actually an old concept. It is like we are being shown the truth is we must get back to basics. In the field of medicine what I am seeing in spades is Hippocrates, "Let food be thy medicine and medicine be thy food." Unfortunately we have been advised of so many things now being shown to be dangerous and we should actually be doing the exact opposite of what we have been told. So, what does this all have to do with energy? Well, this in and of itself does not have much to do with energy but what you will learn in this book will surely help you sift through the recommendations, which have become more and more confusing to decipher. Should I eat low fat or high fat? Is salt good or bad for me? Am I supposed to try to decrease my cholesterol after all?

It has become daunting anymore to know what is truly right for our health and wellbeing. Even the research I conduct on natural healing seems to be indicating we must be very careful of what we follow. For example, since the discovery of antioxidants there has been a huge push for people to purchase antioxidant supplementation. It was all about exogenous antioxidants...until it is now being discovered how some antioxidant supplementation either might be completely ineffective or not suitable for certain people suffering from certain types of conditions.

There is a very popular doctor of natural health who has been promoting his products as well as teaching on various topics. I truly enjoy this doctor's presentations; however, I have become quite overwhelmed by his information. I am the person who spends nearly all day reading or researching various natural healing techniques and modalities yet this doctor literally exhausts me when I listen to him. The problem is not that his information isn't incredibly useful and informative. The problem is he has so much information and so many recommendations I cannot imagine anyone having the ability to achieve the level of health he promotes.

In addition to all of this we know there are many routes toward health using natural methods and it is not a one-size-fits-all approach. In fact what might work for one person might do nothing for another person. While most natural treatments are fairly benign there are certain modes, which might actually hinder a person's healing.

Wouldn't it be nice to know which diet could give us the best results the first time without having to try and fail and try and fail, as is so common these days? How awesome would it be to have the ability to determine whether an opportunity is right without losing money on an unwise decision? Most importantly, what if we were able to self-diagnose and then have the ability to determine the most beneficial course of treatment to achieve ultimate health? I am here to tell you it is possible to achieve all of this and then some simply by tapping into the incredible phenomenon called energy.

Energy is the underlying force, which is utilized by your nervous system and ultimately controlled by the thing we call the "subconscious". In this book I hope to demystify the concept of energy such that you will have an understanding about how it is connected to the spiritual aspect of our lives but in such a way that will prove it is not some new-age hocus pocus concept but actually has roots in science and can ultimately be discussed in scientific terms.

Energy encompasses electromagnetic radiation or oscillation. For purposes of this book energy will be discussed as a broader topic, namely as an invisible and spiritual phenomenon, which has not yet been actually measured, but also as a derivative of electromagnetic radiation and frequency understood on a scientific level.

I decided to write this book first to help the layperson recognize the existence of this concept called energy as a physical manifestation; second to help him obtain a certain level of understanding of energy and its importance; and finally to help bridge the gap between the spiritual and scientific when thinking in terms of energy. During my treatment at the Scher Center for Wellbeing, which will be discussed in a chapter toward the end of this book, I realized, through conversation, how many people have little to no understanding of energy and why it is so powerful when used as part of a treatment modality. This prompted me to write on the subject in the simplest terms possible.

Of course, energy is not a simple concept thus there will be a need for some additional study on the reader's part. It is my intention to present scientific concepts of energy not just as it functions within the body but also how it affects our lives in external ways. Some of the chapters within this book are intended solely to provide evidence and an argument for the existence of energy. Attempting to educate on the topic of energy is no easy feat, especially when my target audience is broad including the layperson – someone who has no scientific background but is nonetheless interested in energy healing – and also for the person who has some understanding of energy, perhaps even a scientific degree, but has not paused to really think about its connection to such concepts as the subconscious mind, intuition, the power of positivity and our connection with all living things.

Because energy is active in all things, especially living things, it plays an important role in our lives thus our failure to recognize it or have some understanding of its function can either place us at a serious disadvantage or, worse, place us in great danger. My goal is to present energetic concepts through a variety of examples to make it not only a plausible idea in your mind but also to firmly establish its existence and power. My goal is not to make you an expert but to give you enough information as to recognize its ability to transform your life.

When thinking of energy in terms of the subconscious mind it is necessary to think outside the box and not expect definitive measurements toward its existence but as something having much power just the same. This being the case there are specific questions we should ask and to which we should seek answers. The questions to ask are: How does energy relate to the subconscious mind and intuition? What *is* the subconscious mind and why is it so powerful? What kinds of things can the subconscious mind accomplish depending on what it "knows"? What kinds of messages make their way into the subconscious? And, finally, is there a way I can control what my subconscious mind receives to ensure the best possible outcome in my life?

In this book these questions will be answered and if you follow the subtle and deliberate suggestions you will not only find better health but also holistic wellbeing in all aspects of your life, including relationships, wealth, and ultimate fulfillment!

Chapter 1

Energy As A Perception

Nothing is accidental and there are no coincidences, especially when it comes to the decisions we make, which shape our lives and experiences. The fact that you are reading this right now is also no coincidence. Something in your subconscious mind led you to this book. Perhaps you've known all along that there is more to life than what we see with our eyes. Perhaps your subconscious mind has shaped each and every one of your experiences in such a way that you were destined to reach this book right here and right now. Some people call this "God," others call it "The Universe." The fact is that our subconscious mind is directly connected to our Source and therefore your destiny could definitely be attributed to the Entity you refer to as your Creator; however, you also have a hand in this, a phenomenon regularly called "free will."

The key to finding ultimate happiness and fulfillment is trusting your subconscious mind to direct you in the best possible direction while using caution with what you allow into your subconscious mind, because it is like a computer storing every bit and byte of information it comes across.

While your conscious mind does not see the details or read between the lines you can bet your subconscious mind most certainly catches – and caches - everything. The phenomenon around which this concept is built is, in fact, energy. We cannot see energy but it exists and we are actually made mostly of energy. So while we can see and feel tangible things they are mostly energy, which can be modified to suit our purposes or even to destroy us. The choice is ours.

As time goes on it becomes more and more clear how relevant, prevalent, and obvious the importance of energy truly is. In order to understand the importance of energy one must first ask, 'what IS energy?' Energy can signify many things but at its core it is in fact the scientific concept of power. Scientifically it can be kinetic (active) or potential. Most engineers and scientists are well aware of the scientific implications of energy. Energy itself is measured in joules, which can be converted to calories, or British Thermal Units (BTUs). Energy is calculated from wavelength and frequency. In scientific terms energy does not equal frequency, however, energy can be discussed in terms of frequency or at least conceptualized in terms of frequency.

Energy encompasses many things but these days we, as individuals, should be more interested in energy as it affects each individual on an emotional, spiritual, and physical level. Although specific studies have not been conducted on energy within the body and how it relates to our choices, health, or ultimate wellbeing, having an understanding of energy as it has been studied scientifically will help to bridge the gap between the scientific and the spiritual. You must have the capacity to envision the commonalities, e.g. the spirit of the body is the spark of life, the energy that allows the nervous system to function. The spirit is the part of a human being that can communicate with and sometimes even see other "spirits". A discussion on frequencies as how they are observed can be found in a later chapter.

With regards to energy as a basic concept it is obvious that society in general is aware of its existence. Many people often talk about various aspects of energy in everyday life. People often discuss the energy of a space, of an individual, and even of an idea. Energy is quite possibly the most important phenomenon today, which has such a huge impact on all of our lives and health. The fact of it being invisible can be very deceiving so as to place us all in various types of great danger. What many people do not know or realize is the potential they have to create not only ultimate health but also even great wealth simply by understanding and harnessing energy. Just because energy cannot be seen does not mean it is not important or, especially, that it cannot help us greatly in our personal lives both internally and externally.

This topic can tend to run the gamut from being über-scientific to über-weird. This book is intended to take some of the mystery out of it so it becomes a viable and worthy topic of discussion with regards to nature's presence all around us and, especially, with regards to natural health and natural healing.

My dog's homeopathic veterinarian, Dr. Dee Blanco, places a great deal of importance on energy in homeopathic medicine. One of the things she says is the main issue that affects dogs [author's comment: and obviously humans as well] is an energetic imbalance. She says we all, as living beings, are mostly energy – "we look much more dense than we are." As difficult as it might be to grasp this concept, anyone with a scientific background will understand how this is possible.

Consider the color spectrum – a most interesting phenomenon when you really think about it. The colors we see are based on reflections. Say what? According to an Internet dictionary color is defined as "the property possessed by an object of producing different sensations on the eye as a result of the way the object reflects or emits light." This means color is not something tangible or even visual but a perception. It is the belief of many, including some of those in the scientific field, that most of our "reality" is a perception as well.

Without venturing into the "Twilight Zone", at least not until it is adequately explained why this is all necessary to understand in terms of "Nature's Presence" and natural health, it is important to recognize how we as a society are far too rooted in what we believe is reality – the tangible and visible.

This is a topic which has been discussed by scholars of all sorts – success coaches and motivational speakers, naturopathic and homeopathic doctors, of course spiritual leaders, priests, etc., and, especially, scientists. It really isn't that weird. We just aren't discussing it enough.

With that said, the next few chapters revolve around some interesting concepts and points to ponder with regards to energy and the things we cannot see but are nonetheless there and affecting our world, our lives, and our health.

Chapter 2

The Importance of a Good Night's Sleep

A very interesting book called *The Body Electric: Electromagnetism and the Foundation of Life* by Robert O. Becker, M.D. and Gary Selden is a must-read for anyone seeking scientific information on energy and how it can affect health and the body. Some of the insights regarding energy discussed in the book you are reading come from *The Body Electric*, as it has provided a highly comprehensive view of energy with regards to health and the human body. However, this book intended to present energy on a more practical level so, if one cares to learn the scientific principles behind the discussion, a read of *The Body Electric* is in order.

Energy is indeed "the spark of life," without which life would not exist. It is responsible for every process within the body, as the root of all bodily functions is the brain. The brain sends electrical signals to various parts of the body through nerves, which communicate with cells about how they are supposed to act. What is really fascinating about this invisible phenomenon is how truly powerful it is. It is so powerful, in fact, that it has the ability to heal or kill, as well as create manifestations of various things, both good and bad.

Over the centuries many people who have shared their knowledge of energy have been ridiculed by the mainstream as being nonsensical. Nikola Tesla was one of the most brilliant scholars to have ever lived but his theories were well beyond the comprehension of the people of his time and thus he was, at times, ridiculed.

It is unclear whether this was because of the disbelief of the mainstream at the viability of his theories or if the mainstream was trying to suppress the truth in the name of helping Big Oil, Big Pharma, or other big business. Whatever the reason, Tesla was always ahead of his time in accepting unorthodox ideas of energy-related phenomena.

For example, hypnotist and healer named Anton Mesmer believed heavenly bodies such as the stars, moon, and planets had direct magnetic influences on earthly bodies but his idea was rooted in astrology, which was anathema in the scientific world in the late eighteenth century. Tesla was the one notable scientist who did not ridicule the theory despite most others shunning the idea.

Over the centuries it has become well known the functions of the earth are heavily in tune with occurrences out in space. For example, oceanic tides are timed to lunar cycles. What many people might not realize is that living beings are also connected with the planets, stars, outer space and, indeed, the universe and not just in an astrological, new age sort of way.

In fact there have been studies conducted that show a direct link between certain behaviors in people and interactions of the sun, moon and planets. The "predictions" so many people rely on in their horoscope might be more rooted in science than we realize.

Like clocks, our bodies are also designed to respond in various ways to solar and lunar cycles. Our circadian rhythms run on roughly a 24-hour oscillation. Various bodily functions are optimized at specific times of the day. It is fairly well known that the most self-healing happens during REM (rapid eye movement) cycles at night while we are asleep. Many people believe sleeping from 9 AM to 5 PM is just as effective as sleeping from 10 PM to 6 AM and thus believe that there is nothing wrong with working a graveyard shift. In actuality those who regularly work such shifts could be causing serious damage to their bodies by depriving themselves of optimal self-healing.

In addition to this connection with solar and lunar cycles other studies have shown that our bodies are synchronized with other geomagnetic activity even to the point of having the ability to predict disturbances in the cosmos simply by recognizing physiological changes as they relate to that activity. Further, scientists have discovered other incredible connections, such as the synchronization of human heart rate with frequencies of the earth.

In one study scientists monitored heart rate variability (HRV) to determine if there were any correlation between this function of the human nervous system and what is termed geomagnetic activity including solar wind speed, solar radio flux, cosmic ray counts, Schumann resonance, the total variations in the magnetic field, and also standard indexes for measuring geomagnetic storm activity.

Before technology became so prevalent people went to bed when the sun set and arose when it came back up again. This was ideal not only for optimizing the workday but also for optimizing necessary bodily functions. Melatonin is involved in the synchronization of circadian rhythms, which are responsible for regulating body functions. It serves many purposes but one of the most important is to facilitate the healing process at night by helping the body fall into a deep sleep.

Anyone with chickens knows they operate on their own internal clock, which is in sync with the sun rising and setting. The chickens will leave their coop at sunrise – and the rooster will crow at that time – and they will go back into their coop at sunset, all without being directed by anyone other than their own instinct to do so.

Due to the pervasiveness of electronic equipment and the exponential increase in its use various disturbing trends have been observed in the animal kingdom. It has been discovered that many insects and birds have a keen sensitivity to the natural electromagnetic frequencies tied to the earth, which greatly affects their sense of direction. When it comes to those flying creatures whose role in the natural ecosystem requires this important function, the disturbance in the electromagnetic spectrum is detrimental to all of life.

In particular there has been a great danger to the ecosystem due to the massive reduction in the bee population, whose role is critical for providing food for all creatures, including humans. It has been postulated that this reduction in bee population is a direct result of the ever-increasing electromagnetic pollution globally. Certain birds, including chickens, have been proven to also have this sensitivity and thus the problem has even further reaching implications, as if a threat to our food supply weren't enough.

Because of these discoveries it has been essentially proven that all of life is connected with these frequencies and thus any disruption in this process is detrimental in unknown and immeasurable ways to all living creatures.

It is believed that the production of melatonin is directly correlated with the radiation emitting from the sun. When the sun goes down and you lay your head to sleep, the brain senses the absence of radiation, which triggers the pineal gland it is time to start producing melatonin, kicking the process of self-healing into full gear. What is likely happening today is that these natural functions have been tampered with due to all of the electromagnetic fields (EMFs) with which we are constantly bombarded.

Along with helping the body to fall asleep, melatonin serves another important role, that of an antioxidant. Even though there might be other, artificial methods, to help the body and brain to fall asleep such as with pharmaceutical sleep aids, they do not have the ability to serve this important function.

It is well known most, or all, illness is caused by oxidative stress in the body and the way this stress is counteracted is with natural antioxidants. Natural antioxidants created by the body are far superior to purported antioxidant supplementation because of the instruction sent from the brain to the cells via nerves, which is based on energy. The brain is embedded with codes, which dictate how the various processes are to occur. Anything unnatural confuses the function by throwing an unknown variable into the equation of these codes and the way they are required to function.

The mitochondria of the cells contain a network of antioxidants consisting of vitamin C, vitamin E, uric acid, lipoic acid, glutathione, and antioxidant enzymes. The body synthesizes some of these antioxidants, such as glutathione; however, others, such as vitamin C and vitamin E are only available through nutrition.

The fact that the pineal gland is located near the center of the brain, in the epithalamus, and is responsible for producing this important antioxidant implicates its importance to brain function. Indeed the brain has its own source of healing, as this antioxidant is reserved solely for the brain and its health.

It is said that melatonin is twice as powerful as vitamin E and five times as powerful as vitamin C. If we are unable to produce melatonin naturally we are placed at serious risk of all illness, as the brain is the command center for all of the body.

It seems the biggest health threat to humanity is cancer thus a great deal of importance is placed on understanding what causes cancer. For this reason working a graveyard shift has been classified as a Class 2 carcinogen, which tells you how important it is for the brain to produce melatonin.

Because the pineal gland is triggered when it senses the absence of light from the sun, when the sun sets the brain has no idea it must produce melatonin because it is still sensing all of the other radiation and EMFs in the environment. This includes EMFs coming from the electrical wiring in the walls, the high definition television in the bedroom, the appliances in the kitchen and throughout, the wireless router communicating with the devices in the house, the smart meter placed on or near the house and, this is a big one, your cell phone next to your bed.

Thus ensuring that you get ample sleep during the night and "quieting" your sleep space is critical to optimal healing. For this reason I turn off my wireless router at night as much as possible and ensure my cell phone is in another room, far enough from my head so as not to cause much damage. I am unable to practically shut off all electronics near me, such as the wiring in my bedroom, but we would do well to limit such things as much as possible so we could attempt to have a healing, restful, night's sleep.

Chapter 3

Electromagnetic Pollution and Health

"Except for light and infrared heat, we can't perceive any of these energies without instruments, so most people don't realize how drastically and abruptly we've changed the electromagnetic environment in just one century."
-Dr. Robert O. Becker

Energy deriving from electromagnetic waves is expressed in the unit of measure called Hertz (Hz) or cycles per second, a measure of frequency. Essentially it is the frequency of oscillations or vibrations that is measured. Frequency is in all things, as they exist in solid form at various frequencies. We are able to feel because of certain frequency ranges, and we are able to see and hear because of certain frequency ranges. All things emit electromagnetic energy at various frequencies. Electromagnetic energy exists at various frequencies ranging from microwaves at billions or millions of cycles per second (gigahertz or megahertz) and even higher to radio frequencies (RF) down to extremely low frequencies (ELF) measured in microhertz (μHz) and even lower. In between are the UHF (ultrahigh frequencies), VHF (very high frequencies), etc. With the increase in electronic technologies there has been a great increase in these frequencies, most of which are unseen.

The natural frequency of living beings is fairly low and generally energy is only absorbed in those ranges, with only a small amount of energy absorbed at higher frequencies. Prior to the electronic age this meant living beings were exposed to an energy spectrum comfortably adapted to and which did not cause major issues physiologically, psychologically, emotionally, etc. The main sources of energy were in the cosmos and the low, naturally occurring and healthy fields of the earth. With the advent of everything from battery operated devices to metal refining plants, antennas, cell phone towers, electric substations, TVs, satellites and beyond, humankind and indeed all living things have been subjected to electromagnetic frequencies.

Energy emitted in the background by every array of substation, electronic meter, and transmitting device imaginable is called electromagnetic pollution. If this type of pollution were visible to the naked eye it would look like the heavy smog seen in some of the largest, most polluted cities. Although electromagnetic pollution is not seen, just as visible smog can cause health problems, it is invisibly wreaking havoc on our bodies nonetheless.

Some of the damage caused by EMFs is so subtle we barely realize it is happening. It can manifest as stiffness in the joints first thing in the morning, headaches at various times of day, and a lack of sleep at certain times of night. The main issue is that the human body synchronizes itself to certain types of energy, as all living beings are connected. Many artificial energy sources, such as computers, cellular telephones, and other electronic devices, for example, are far enough outside the normal human or living energy spectrum that the body is unable to synchronize, which causes disruption in many bodily functions.

Even worse, certain environments can increase the damage several-fold. For example, rain creates reflective surfaces allowing these signals not only to pass through the environment but also to bounce off surfaces creating an increase in the amount of signals. These rogue signals cause damage on a cellular level and also by disrupting the communication between the brain and the body through the nervous system.

Unfortunately unless we band together to demand a reduction in these EMFs, we are stuck with this pollution. Things will not be getting better anytime soon. The question is what can we do on an individual level to protect our families and ourselves so we are disease-free or at least minimizing the damage from this pollution?

First, unplug as much as possible. This means minimizing your viewing of TV, computer use, and cell phone use. It is never a good idea to have a TV in your bedroom, as it is too tempting to keep it on while you sleep, which further disrupts the healing process of the body. Needless to say the TV in your room is causing other types of damage as well, in the form of subconscious or subliminal messages, which will be discussed further in a later chapter.

Unplug other devices as well. If you are not using your appliances, unplug them because even when they are not in use they are emitting energy as long as they are connected to an electrical source. By no means should you sleep near any electronic devices; even simple digital alarm clocks are not a good idea. The worse thing to do is to sleep next to your cell phone, especially if it is receiving wireless signal.

Turn off your Wi-Fi router at night and when it is not in use.

Finally, learn to use techniques that help clear the EMFs in your own space. Because we are energetic beings and our brains are capable of sending its own signals to our bodies, we are capable of doing much more than we realize to affect our electromagnetic environment.

Chapter 4

The Implications of Cell Phone Use

For several years now, experts in the field of electronics as well as popular natural doctors have been discussing the dangers of cell phone use. Articles written by these scholars are quite disturbing to say the least. It has been surmised that one day the masses would realize cell phone use as being more dangerous than smoking cigarettes, which was totally accepted at one time.

An article that came out in the summer of 2013, written by William Thomas titled *Wireless, Chemtrails, and You*, goes into great detail regarding the dangers of wireless technology and the invisible frequencies to which our bodies are unaccustomed. That article discussed not only cell phone use but also other types of technology such as "Smart Meters" used by electric companies. It is highly recommended to research "cell phone dangers" and "smart meter dangers". The results will prove to be quite shocking, no pun intended.

Epidemiologist and medical scientist George Carlo, Ph.D., J.D., was commissioned by the wireless industry to conduct safety studies between 1993 and 1999. Of course the wireless industry was expecting for their scientist to proclaim the technology to be harmless. What actually happened was that Dr. Carlo could not make such a proclamation and ultimately did not accept the industry's money in favor of having a clear conscience.

The major problem with cell phones is they emit microwave radiation and it is most concentrated near the surface, thus, when the phone is near the body, it is constantly irradiating the person. Some studies have been conducted on their dangers and although these studies have proven the dangers, because the industry is very profitable the results are suppressed. Outwardly the industry claims the only dangers are related to the heat created, which can mimic microwave energy, and since they do not get hot enough to do any real damage to cell structure they are deemed "safe." The truth appears to be much more sinister.

Studies have shown that the radiation emitted by cell phones, which is in the radiofrequency (RF) range, can cause cancer and tumors by penetrating the skin and creating changes in the cells. Anecdotal evidence has shown women who tuck their cell phones in their bras to have an increased rate of breast cancer and men who wear their cell phones on their hips to have an increased rate of prostate and testicular cancer.

The dangers of using cell phones as intended, by placing them next to the ear, have far reaching implications because they have been shown to penetrate the blood-brain barrier causing leakages, killing brain cells, and creating tumors. Of course cancer is likely not so simple and our bodies can withstand quite a bit. Those anecdotal cases could have been a result of the perfect Molotov cocktail of issues that culminated in a tumor.

Indeed, brain tumor diagnosis is on the rise and since it has been shown that cell phone use can cause gliomas and other dangerous brain tumors, coupled with the fact that the use of cell phones continues to increase yearly, it would appear a correlation could be made with regards to this alone. Unfortunately, because scientists have purportedly not pinpointed the actual mechanism of how cell phone use can cause cancer, the industry has avoided ramifications.

Due to the "everyone is doing it, so it must be okay" attitude, the general public fails to recognize how serious the issue truly is. Despite its failure to publicly educate the masses on the dangers of cell phone use the wireless industry is actually well aware of the problem. The paper insert found with most, or all, smart phones provides recommended distances between the device and the body as well as warnings and other disclosures. It is unlikely many people read this information. Because it is also difficult to locate, and is purposely buried within the fine print, clearly the wireless industry seeks to protect itself from liability while ensuring continued profits.

Whether the use of cell phones directly causes tumors, or even if it just contributes as part of a combination of dangerous habits, it is clear we should minimize our exposure to anything that could contribute to such a dangerous cocktail. Conducting a meaningful search on the subject will put a fright in anyone, and it should. The implications of cell phone use is beyond disturbing and the fact that this technology has been foisted upon the general public before any studies were conducted, essentially using the people as guinea pigs, is irreprehensible.

Cell phone use with regards to children and the next generation might be the most disturbing. The increase in wireless technologies is thought to be responsible for a major decline in fertility among couples in first world countries. "At current rates, plummeting global sperm count will hit 'zero' by 2048". William Thomas, *Wireless, Chemtrails, and You.* Another disturbing trend is the increasing number of children with access to cell phones at a young age. Well meaning parents who allow their young children, who have obviously not yet fully developed, to use wireless devices may be inadvertently shortening the lives and creating a potentially dangerous health situation for their beloved children.

The younger the child the more dangerous it is for him to handle wireless devices or even to be near them. It is not just a matter of not allowing them to talk on an activated phone but even just holding the device, such as for playing games, as it is sending and receiving signal, can affect the still-developing delicate cells in the child's body. As a child talks on a cell phone with it placed next to his head the radiation is penetrating the child's malleable skull and blood-brain barrier, which are still very thin, as they have not fully developed. It is imperative not to allow children to be near wireless devices as long as they are sending and receiving signal.

It has been found that talking on a cell phone even two minutes per month can cause brain leakages creating breeches of the blood-brain barrier, potentially causing major brain illness and cancer. It has also been found that the use of a cell phone for ten years by anyone increases his or her chance of getting cancer by 500%. The problem is not just cell phones either. As previously discussed, all types of wireless devices are creating the same cellular changes in the body, even wireless handsets connected to a landline.

In truth, many scientists do believe they understand the mechanism of cancer and tumor development as a result of wireless technologies and it goes back to getting a good night's sleep. As discussed in this book, melatonin is produced by the pineal gland during sleep and, in particular, when the body detects an absence of light, or radiation. One of melatonin's important functions is to act as an antioxidant. So, even if one gets enough "sleep" but his sleep environment is filled with electromagnetic pollution, including and maybe especially sleeping next to the cell phone the pineal gland is likely not triggered to produce melatonin. The body's self-healing function is not just a matter of being asleep but of having important processes within the body functioning optimally. One of the most important processes is the body's ability to fight disease through addressing oxidative stress with natural antioxidants.

Unfortunately the wireless industry continues to feed its addiction to profits as people feed their addictions to fancy devices, seeking faster speeds and stronger connectivity and therefore the industry continually seeks to "improve" the network. Currently the fifth generation of wireless technology, or "5G", is beginning to unfold. Purportedly this new technology will not require cell towers but can transmit signals from boxes that are to be placed just about anywhere in every single neighborhood. The powers-that-be are claiming it is not a matter of "if" this happens, but when.

In addition to the incessant "improvement" of the wireless network there has been an increase in "smart home" technologies, which utilize devices such as smart meters touted by electric companies as being beneficial toward energy conservation but which actually seem to be more of a control mechanism and a form of "big brother." Much like living energetic beings can synchronize with one another, it has been said that, at certain times of night, smart meters synchronize with each other as a form of communication and many people become aware of this synchronization by waking up in the middle of the night and struggling to fall back to sleep until the communication process is complete. Imagine the disruption occurring due to this alone.

It is believed by many scientists that our ecosystem, which consists of a specific and delicate balance of plants, animals, and humans, is hanging on by a thread. It would seem the greatest threat to humanity is the ever-increasing obsession with technology and control along with the geoengineering, which is presently being conducted purportedly to enhance certain technologies and to shove humankind into some sort of "evolution" toward becoming humanoids.

I realize this sounds a bit off the wall; however, if you conduct a search into geoengineering with an open mind you will see many ideas and theories relating to the constant quest of first world countries to have the latest and greatest technologies not only for profit but as weaponry as well as a means to control the populace. Invisible energetic weapons are a serious threat and we should all be aware of how to protect ourselves from such technology if possible.

So, what now? As long as society continues feeding its addiction to "devices" creating a profit for large corporations there is not much we can do about the wireless use around us. It would be nice if the majority of people would become active in their communities to refuse the 5G Network. This technology has the potential of being the tipping point toward total disruption of natural electromagnetic frequencies within the body and in other living things, as well as the earth itself. The prospect is indeed terrifying.

Beyond a miracle of getting the majority of the population involved we can, however, affect change in our own households. Limit the use of cell phones as much as possible. Turn off or deactivate wireless devices at night while sleeping. I realize I have mentioned this before but it cannot be emphasized enough. Also, place devices in airplane mode when stored near the body. And by no means should children *ever* be allowed near an activated device.

As much as possible become active in your community to educate others on these dangers, especially your elected officials who might have some ability to effect change. Also, become more educated with regards to the smart meter technology and look toward the policies of your local electric company on how to opt out of the use of these devices.

With all of this said, believe it or not, even this type of energy can be controlled using certain tools and shields as well as the power of thought and understanding how energy functions within our bodies. To understand more about this amazing phenomenon, read on.

Chapter 5

Energy and Regeneration

Energy, as it manifests in different forms, is essential to every form of life. Plants and many species in the animal kingdom require energy for survival in terms of regeneration. Regeneration is an elusive phenomenon many people do not understand and the scientific community has largely ignored but the few comprehensive scientific studies done on regeneration have yielded some interesting results.

In order to understand the mechanism of regeneration scientists have studied animals that are able to either regenerate nearly completely or even just minimally, such as lizards, which, as anyone who had a healthy childhood knows, can regenerate their tails after they have fallen off. As it turns out salamanders and newts are able to regenerate nearly their entire bodies, including their little hearts even, or especially, on the brink of death. Using this incredible information scientists have attempted to understand this mechanism so they could possibly use it in humans, particularly those who have lost limbs.

Regeneration is the important process of the body's ability to regrow itself. Humans enjoy a very limited amount of regeneration in the form of a debilitating liver, which, if it is not too far gone, can regenerate, and also in the form of bones, which regenerate to fuse back together. What is not known is, with further research, humans might be able to regenerate entire limbs depending on the location of amputation.

It turns out that with the help of certain frequencies of electricity such regeneration could be possible. Scientists have determined the driving force behind regeneration to be the electromagnetic impulses sent from the brain through the nerves to the bone. They have been able to accelerate regeneration in certain animals such as salamanders and frogs by enhancing those frequencies to a certain extent. Prior to these experiments it was believed only certain reptiles were capable of regeneration but due to these studies further research was done early on to attempt to determine its viability in humans.

In the early 1970s one "poor" child became the inadvertent subject of an experiment on human regeneration of bones. This child had lost the tip of her finger and, due to some sort of red-tape mishap, was unable to see a surgeon soon enough to have the missing portion replaced. Due to this failure the fingertip of the child regrew perfectly including a perfect nail bed and fingernail!

Numerous other situations of many children under the age of eleven who had been "neglected" and also actually regrew their fingertips have been documented. This means humans have a certain amount of regeneration ability, at least under a certain age, indicating such process could potentially be artificially simulated in adults. Unfortunately the scientific community has largely ignored such amazing findings and has failed to conduct further experimentation on the "phenomenon" to determine how it could be effectively used in amputee patients.

The point of this illustration is to show how energy plays such an incredible role in our bodies despite our inability to see it happening. We should no longer ignore it as a viable force to reckon with, no pun intended. Its invisible nature should not and does not mean we should not be aware of its place in our world, namely because there are many healing modalities involving energy and by recognizing its potential, literally, we each have the power to heal ourselves with our own energies – even possibly to the point of regeneration.

Chapter 6

Energy in Acupuncture

It is not difficult to find many people who swear by the efficacy of acupuncture. On a personal level I have found acupuncture helps with muscle tension relief and very general issues. For others it has helped with everything from digestive issues to female health issues.

At the time I was being treated for allergies the acupuncturist made some interesting statements and had some interesting insights into my health. One observation made was about my "Qi" (or Ch'i), which the Chinese believe to be the energy (life) force of the body. She told me my qi was weak. This was news to me although I did have a few issues I assumed were just part of life – allergies as well as digestive issues, which, incidentally, she was able to pinpoint as well by gently massaging my legs and asking if they were tender. The answer was a resounding yes; they always seemed to be tender to the touch. It turned out this was a sign of digestive issues.

Around the same time, in the parallel universe of mainstream medicine, I had been diagnosed with and was being treated for Acid Reflux.

When I asked why she said my qi was weak she stated it was because my body was not providing a stable medium for the needle; it was like "inserting the needle into tofu." This was slightly disturbing and prompted me to take more control of my health by eating better and attempting to deal with stress in a healthier way.

The more one studies this phenomenon of energy the easier it becomes to conceptualize things such as the acupuncturist's statement of my body being "like tofu." Considering our bodies are made up of a conglomeration of molecules "glued" together with energy then the analogy of tofu as indicating a weak qi or energy force would stand to reason. If we can see energy as the glue that holds our cells together then, if the life/energy force is weak, so would be the glue and the molecular structure of the body would not be as tightly compact as it should or could be.

The premise of acupuncture lies with the meridians in the body through which the life force/qi passes allowing for optimal health and vitality. If any of these energy pathways is blocked various issues ensue. Using needles acupuncturists either clear the pathway or redirect the energy so it flows continuously. Occasionally an acupuncturist will use heat, pressure or slight electrical currents to aid in clearing the blocked path. Thinking about the function of energy in the body it would seem that by clearing blocked energy pathways, the brain's messages of healing or cellular instruction through nerve impulses would be much more effective.

Many people use acupuncture for pain relief. The body feels pain due to impulses sent by the brain to various locations in the body. The way pharmaceutical pain relievers work is by interrupting certain receptors and "bypassing" the brain's natural reaction to pain. This is detrimental for various reasons, one of which is that our body feels pain to get our attention. Pain is the body's way of telling us to pay attention, something is wrong. Moreover, blocking certain receptors in the brain for the purpose of pain relief or anti-anxiety could put one in a dangerous position in the event of an actual threat. Pain is also a way to alert the body of impending danger.

Blocking the pain might provide relief in the short term but it could mask something very serious, which should be dealt with. Further, pharmaceutical pain relievers contain toxic and often very dangerous chemicals, which are not only addictive but could permanently alter brain chemistry. It seems then that a modality such as acupuncture would be a good alternative to pain relief, as it uses the body's very own energy pathways to create a "natural" pain relief. In the meantime, one who suffers chronic pain should incorporate other treatments to deal with the underlying issue.

It is always good to research providers so you know you are receiving the highest quality of care available. It is recommended to find an experienced acupuncturist who is also willing to explain the concept to his or her patients so some of the "mystery" of acupuncture is removed.

Chapter 7

Energy in Homeopathy

"We are mostly energy; we look much more dense than we are."
-Dr. Dee Blanco, Homeopathic Veterinarian

Homeopathy is well known as a viable alternative to conventional treatments. While the general concept of homeopathy is "like cures like," it should be understood there are important energetic aspects to homeopathy. Because this is a discussion on energy and is the focus topic it is recommended for you to study homeopathy more deeply to better understand it as an effective treatment modality. For purposes of this discussion, however, the short explanation is homeopathy is comprised of using the same substances or essences of the substances, which cause certain, specific, symptoms to cure the very same symptoms.

Treatment is done using very diluted quantities of the substance and administering it in a specific way. The substances used in homeopathic treatment vary greatly and come from numerous sources such as plant essences, including poisonous plants, to diseased organs of animals.

Samuel Hahnemann, the doctor who is considered in many circles the "father" of homeopathy, proposed that the essence of illness is a disorder in the vital force; the vital force being the invisible energy discussed in this book. By taking a holistic view of the patient's symptoms and other signs, such as behaviors, a homeopath could see a picture of the vital force even though the vital force itself cannot be seen.

Homeopathic remedies are prepared through a series of dilutions. The doses of homeopathic medicines are greatly diluted to ensure no toxicity whatsoever and, really, no traces of the substance at all. The higher the number of potency of the medicine the more it has been diluted. Potency is self-explanatory. The more potent the medicine it would seem the quicker it would work or the more effective it would be. Of course homeopaths have a reason to prescribe lower potencies, as "more is not always better." For the purpose of this discussion it is necessary to attempt to get to the bottom of this to try and understand how energy comes into play with homeopathy.

As a higher potency means greater dilution it thus means that, statistically, no atoms or molecules from the original substance are left. This would seem very counterintuitive as to the efficacy of these homeopathic remedies if we view those remedies as having value due to the substance itself.

In fact, it has been hypothesized that the way homeopathic remedies work is not so much due to the substance itself, which causes symptoms in healthy people, but the energy of the substance that triggers the body's own self-healing properties and thus cures the issue. This would mean it is not the chemical structure or molecules of the substance facilitating healing but the energy of the substance, which is left behind.

It is said water holds an imprinted memory of everything to which it has been exposed. The theory is the water in which the substance is diluted holds the memory of the energetic nature of the substance therefore none of the molecules are necessary to facilitate healing. Thus is the nature of energy. It can be embedded without any trace of physical molecules present.

Although it is not clear why, it appears acupuncture sometimes does not mix well with homeopathy. Chinese medicine is largely rooted in energy healing therefore it has been found also that other Chinese treatments such as those with moxa (heat) and Chinese herbs also do not mix well with homeopathy.

Many natural doctors who practice both acupuncture and homeopathy have found it is most effective to use one or the other on a patient as opposed to both at the same time even though other alternative treatments, including oriental massage, do not interfere with homeopathy. Perhaps the missing link to solve this mystery is this energetic connection that resides in both acupuncture and homeopathy.

When it comes to homeopathic remedies the energies from the various substances are diluted and the remedies are placed onto sugar pellets. Homeopathic sugar pellets are administered by either placing them directly under the tongue or dissolving them in water or some other neutral fluid for absorption in the mucous membrane of the mouth.

To make them effective when administering via fluid, several pellets are placed in a bottle containing the water or neutral fluid (wherein water is the base) and succussed. Succussion means shaking the solution vigorously. It is not necessary that the pellets are completely dissolved but the succussion is necessary because this is how the energetic nature of the remedy is activated. In many ways administering the remedy using this method is much more effective than simply dissolving the pellet under the tongue.

A popular medicine that has been used for decades in patients with heart failure or heart disease is nitroglycerine. Many people do not realize that nitroglycerine is a homeopathic remedy, which can normalize blood pressure and even prevent a heart attack. It is administered by placing a pill or two under the tongue. I remember my own grandmother, who suffered from heart disease, always had nitroglycerine tablets on hand.

Anyone who has used homeopathic remedies as directed knows you are not supposed to touch the pills/pellets with your fingers or skin or it will make them ineffective. The pills/pellets should be placed directly from the vial to the mouth, i.e. under the tongue. Neither should the tip of a dropper bottle containing the remedy in liquid form touch any part of the body.

Something to ponder is why touching the medicine with skin would deactivate the remedy and how this direct administration of the medicine would ensure the energetic properties remain intact. Could it be something as simple as the oils in the fingers or hands convoluting the remedy as the reason for deactivation or could it be that, by touching the remedies, we are transferring our own energies or the energies of other things we have touched thus deactivating the energetic component of the remedy?

It has been also found that these energy essences, which are present in homeopathic medicines, can also be transferred through other media, as discussed in a later chapter. Many modalities use this concept to effect healing.

Chapter 8

The Power of Thought
and the Law of Attraction

This chapter introduces the concept of the law of attraction with a personal story, which was only apparent later as being attributed to the law of attraction. It is incredibly important to understand this phenomenon, as it has the ability to help one manifest many things large and small, including health, a massive fortune, fame, illness, or even death. Indeed it cannot be stressed enough that you hold within you incredible power, a power that can make or break you.

In the spring of 2014 I had a nice long chat with a good friend of mine from Carlsbad, California on the subject of energy healing. I met her because of her brother, my favorite author. The story of how we became part of each other's lives deserves some recognition as to the premise of this book. It is a testament to the power of thought and the "law of attraction," both of which are attributed to energy.

In 2005 I was browsing Amazon when a book suggestion for *Burro Genius* came up. Mexican-American author, Victor Villaseñor, who had written the New York

Times Best-Seller *Rain of Gold*, was the writer. I was not familiar with any of his work at this point. The memoir of his upbringing in Oceanside, California looked interesting so I ordered it. Once I began reading the book I could not put it down. The story was so interesting and so vivid, I was compelled to order most of his other works and a few copies of *Burro Genius* and *Rain of Gold* to give as gifts that year for Christmas. He soon became my favorite author simply because I loved all of his books, which are mostly memoirs and biographies.

Around 2007 my husband, Gil, and I were making a daily 3-hour commute to and from work. I decided "we" would read *Burro Genius* again to occupy our commute. It quickly became a favorite of his as well and reminded me of how I had wanted to try to get Victor out to my local community for a speaking engagement the first time I had read the book. This is when I met his lovely sister, Linda. She was immediately welcoming and we were fast friends.

In *Burro Genius* Victor vividly describes the magnificent rancho his father built back in the 1940s. He talks about the parties his parents hosted on the grounds and paints a very clear picture of this beautiful place near the Pacific Coast. In addition to my desire to bring him out for a speaking engagement I placed it firmly in my mind I wanted to visit his rancho and see it for myself. How this would happen was beyond me at the time but I decided it would happen. And it did but it took a little while.

In the meantime, as the board president of a local non-profit, I managed to secure him for that speaking engagement in my town. He spent time with my husband and me at our house as well as our being his tour guide for a little bit of sightseeing. I learned quite a bit from Victor during his trip as well as subsequent telephone calls.

Although his beliefs struck me as a bit "new-agey" he helped me seek a deeper wisdom and understanding, particularly with regards to the energy written about in this book.

Linda and I continued our friendship and spoke with each other occasionally. In 2009 Gil and I embarked on a two-week motorcycle ride to California with the intention of riding along the Pacific Coast Highway, the 101. Our first stop was Sedona, Arizona and our second stop Oceanside. Linda had arranged for Gil and me to stay at the Rancho with Victor. Needless to say it was a wonderful and whirlwind trip. We had intended on staying only one night in Oceanside but we enjoyed it so much, both the location and their hospitality, that we decided on staying a second night.

It was not lost on me how I had made a firm decision, *years before*, that I *would* visit that Rancho some day, and it happened. I had read the book *Think and Grow Rich*, by Napoleon Hill, which discussed this very thing, but I did not really believe it until I experienced it myself. Our minds are very powerful; they can help us manifest many things, both positive and negative. Being positive and having faith is a large part of making our dreams come true...and it is all related to energy.

Not only did the law of attraction allow me to fulfill my goal of visiting Victor's rancho but it also led me to his sister Linda, who shares an affinity for natural health and, in particular energy healing. In fact it was Linda who directed me toward Innerwise, a critical component of this book and the explanation of energy as it relates to health.

Chapter 9

The Power of Thought:
Think and Grow Rich

One of the best and most popular books on the concept of the law of attraction has been around for many decades, nearly a century. *Think and Grow Rich*, by Napoleon Hill, was published in 1937 and is still relevant today, with recent famous motivational speakers still referencing it in their work. The book must be read with an open mind and heart and with the knowledge you have received from the book you are now reading. Although it might seem like it is rooted in new age concepts, it is actually well rooted in science. Mr. Hill talks not only about how powerful our thoughts are and the fact that they can attract or repel certain things from our space but also about the energies associated with the law of attraction.

Once you expose yourself to various books on the subject, and read them with the idea that this phenomenon can be proven through science, you will be able to grasp concepts that once might have seemed outrageous. *Think and Grow Rich* goes into much detail about how our thoughts shape our reality and how we can manifest certain things in our lives simply by thinking a certain way and tapping into the subconscious mind. It also discusses how energy relates to our ultimate lot in life. Other books on the subject are very similar in promoting positive thinking and steering our thoughts to achieve specific goals.

Think about those days when you wake up "on the wrong side of the bed" where you start the day stubbing your toe and you are so upset it is all you can think about. Following this you step into the shower and find the water is cold. This upsets you as well so you dwell on this. Before you know it your entire day is shot and you can't figure out why. The theory is that you helped to create your bad day by dwelling on the negative, which started with stubbing your toe. It is almost as if you expect to continue having a bad day. If you had just forgotten about it, put some positive thoughts into your mind, or even viewed the "negative" event as something to facilitate the positive, and moved on you might have turned your day around immediately. The good news is this can be done at any time throughout the day or your life.

Sometimes it takes something tangible to happen in order for you to believe in this concept. Many times there *are* tangible things that happen but if we are not looking for them or at least noticing them we miss the lesson. Even very minor things are the result of the law of attraction. Sometimes we have to think back and remember to the time when we were hoping for this or that thing to happen when we finally realize the power in thinking.

It took a few years for me to achieve my goal of meeting Victor and seeing his rancho. If I were not aware of this concept I might not have recognized my internal power to make things happen. It is not to say this is an easy task. In fact it is quite difficult, especially when it comes to certain things like actually getting rich. You must work toward what you seek to achieve and ensure you have the proper mindset. Becoming rich is one area in which you must practice consistently. However, ensuring the concept is firmly rooted in your subconscious mind makes the desired outcome much more likely to happen.

In truth there are many examples in my own life that exemplify the concept of the law of attraction. It happens when you run into someone you haven't seen in a while but about whom you have been thinking or talking. Once it happened in a negative way when one morning on my way to work I thought, "I haven't been in a car accident for quite a while." You might imagine what happened. Later that day, as I was going through a traffic light on the right lane, an elderly lady attempting to make a right turn from the left lane sideswiped but just about t-boned me. Let's not dwell on that idea. However, it illustrates how incredible the power of thought truly is.

On a spiritual level this concept is called faith. I have always had a great deal of faith, thanks to my parents. They always said, "God will provide" and I always believed it. Not surprisingly there have been times in my life, particularly in college, where I would look back on a previous month and the funds that were – or were not – available and wonder how I even made it through that month. Of course these are also the types of thoughts not to dwell on but it is an interesting experiment.

In scientific terms this concept is called energy, the subject of this book. According to Dr. Becker from *The Body Electric*, "We know, on the psychological level, that a person's emotions affect the efficiency of healing and the level of pain, and there's every reason to believe that emotions, on the physiological level, have their effect by modulating the current that directly controls the pain and healing." Fascinating!

This is perhaps the reason we hear of people who are diagnosed with cancer and given a very short amount of time to live but actually survive even decades. It has been a theory of mine that those who trust doctors without question and are given this type of prognosis actually only live the amount of time given by the doctor. Obviously doctors must protect themselves legally by being "realistic" but I believe it is these types of prognoses that can shape a patient's thoughts and thrust him into the negative so instead of "thinking and growing healthy" they "think and grow sick."

It seems to be when the patient takes this type of information with a grain of salt and has a healthy attitude as well as faith the patient indeed heals and lives many years. A friend once told me this very thing about her father who had been given 6 months to live. He refused chemotherapy and radiation and told the doctor he wouldn't be back to see him. He recovered and was healthy 23 years later! It is akin to the concept of "spontaneous healing", which really is a thing. Even a patient undergoing toxic treatment is capable of healing simply due to his faith and the power of thought. It is rare but such people, whose faith is so strong it can overcome the most horrific of tragedy or adversity, do exist.

Ponder on how your thoughts have manifested into reality and use it as inspiration to think and grow in a positive way.

Chapter 10

The Power of Thought:
Law of Attraction Take Three

An image with the following quote was circulating on Facebook recently and it was timely for this book. It states, "Everything is energy, energy follows thought, thought becomes belief, belief determines your reality, and reality shapes your destiny!" Think about the concept of "luck" and how it relates to "The Law of Attraction." Some people insist they have "bad luck" and this is why they cannot get ahead or why they are always stressed out. Even when it is not referred to as bad luck there are certain "types" of people who simply believe their destiny is to be under stress.

Some people believe they are cursed with bad luck. We have all heard, "If it weren't for bad luck [he/she] would have no luck at all". This is such a debilitating idea people don't realize how much harm they are doing to themselves by both saying this and, especially, thinking it. By no means should you ever doom yourself to bad luck by expecting it to happen. This is the law of attraction at its worst.

One way to overcome resistance to the idea that we become what we think is by taking personal responsibility.

Taking responsibility necessarily results in one delving into the reasons for, in particular, the "bad" things in one's life. We are unable to blame these things on others and, as a result, we tend to seek the truth, which inevitably leads us to the concept of the law of attraction. Blaming such things on "luck" is also nothing more than victimhood and a failure to take responsibility.

It is not easy to take responsibility or change our thoughts or attitude. Often it takes some major adversity such as a death, serious illness, home foreclosure, or divorce to force one into change. These situations can truly make or break a person. Many people struggle with these things feeling completely defeated and desolate even going so far as blaming God for their problems or losses and even "hating" Him. This is a heartbreaking concept, as it is in doing this that one squanders the opportunity for great change and great possibilities. Even worse, just as with the "bad luck" theory these people will live various individual nightmares over and over again because they fail to see the opportunity in the adversity and the beautiful lessons to be learned while instead dwelling on the negative.

Again, this is not to say it is ever easy but it is simple to do and it is a simple concept. It is extremely difficult not only to see the opportunity but also to pull oneself up out of the desolation and make changes, especially internal emotional and psychological changes; however, it is simple to steer thoughts from the negative to the positive. One of the reasons it is so difficult to actually effectuate change is because it takes courage to look into oneself and see flaws.

We hate to admit we are not perfect, even though we know we are not. It is so much easier to be a victim. To take responsibility is to admit we have flaws, which is to admit that others might have been right about us whenever we have had conflicts. Not only this but taking responsibility is to avoid blaming others or other things for our destiny and placing accountability squarely on our own shoulders. It means being in control and releasing any feelings of being powerless. To overcome the victim mentality is a beautiful and incredibly liberating thing.

This is where the power of thought comes in and how energy can be used to help a person change his or her mindset. It takes some training and effort especially when you are accustomed to dwelling on the negative. It takes even more effort when you are surrounded by those who like to dwell on the negative or are accustomed to sympathizing and feeling sorry for others. The victim mentality is crippling.

None of this is to say that I believe we necessarily create our own reality. Indeed it almost seems that way but that idea would undermine the many choices we have on a daily basis. Just because we set an intention of being happy doesn't necessarily mean that the happiness we ultimately achieve is that which we had envisioned. Besides, happiness itself can only be had through the journey. Ultimately it is joy that we seek.

I recently listened to an interview between a natural health radio host and a gentleman who cured his own Lyme disease using only herbs. Apparently this man is also an expert in affirmations. He discussed how one could achieve or acquire anything they wish simply by verbalizing or, more importantly, writing down their desires. What he said was that it is important to state affirmations in the first, second, and third person. Stating the affirmation in the first person breaks the negative statements we regularly place upon ourselves; stating it in the second person breaks the negativity that comes from others directed at us; and stating it in the third person breaks the negative projections from others upon us indirectly.

What he had to say was most fascinating. Being well versed in this practice he explained how he has "manifested" income in his mailbox within just a couple of days. He also mentioned the story of a man who has attributed his great wealth to his writing of affirmations down every morning, which he does religiously. Purportedly he has seven affirmations he always uses.

The fact that these men actually achieve what they speak out loud and write down makes one think we have the ability to create our reality; however, one thing he said struck my curiosity. He said that one of the affirmations should always be, "I am willing to accept whatever I receive that I ask for." To me this says we do not actually create our own reality and, although we have the power to manifest whatever we wish or whatever we envision – good or bad, unfortunately, how we get there can take many unexpected twists and turns. Could this be the basis of the common saying, "Be careful what you wish for"?

This being said, happiness is most certainly a journey rather than a destination. You will have setbacks and faltering in your thinking; you will slip into victimhood at times forgetting this advice. You will find you have created a negative situation by dwelling on unsavory thoughts. Although some people are quite remarkable in turning things around it is only human to fall into old habits. This is why the best advice is to always be present wherever you are. Remember each moment is special and within each moment is an opportunity. Rest assured that there is always a light at the end of the tunnel and the length of that tunnel is all up to you.

Chapter 11

Energy Healing: Innerwise Part One
The Importance of Energy Healing and Spherical Vision

For the purpose of this book and conveying a logical explanation of the concept of energy as it relates to the body and to health I would like to highlight the healing program rooted in energy healing with which I am most familiar. Although there are other such programs and modalities, I feel this particular program provides great insight into the science behind the mystery of energy within the body.

Believe it or not many people, including medical doctors, around the world have a great understanding of the energy within the body and energy healing. In fact it is much more mainstream in other countries than in the United States. One particular practitioner of a treatment modality surrounding energy healing is a gentleman named Uwe Albrecht M.D., a German medical doctor who developed a natural energy healing technique called Innerwise. With Innerwise the individual uses his or her own intuition and body as the diagnostic tool and energetic frequencies as the healing mechanism.

Through mentors Dr. Albrecht gives workshops, which consist of discussion and practical training. In a two-day workshop he explains the science behind Innerwise along with practice. Once a person completes the workshop he or she receives a certificate of completion.

The web site gives a great deal of information regarding how it works and, once you become familiar with it, you might see similarities between Innerwise and treatment modalities which use kinesiology. A local chiropractor, Dr. Gary Storkan, the doctor who wrote the foreword of this book, has gained a large following due to his ability to help cure allergies noninvasively. The premise goes back to the discussion of the subconscious mind knowing exactly what is going on with the body and can communicate those issues through imbalances in our bodies. Dr. Storkan himself clears energy blockages using physical tools whereas Dr. Albrecht uses energetic frequencies.

The website, http://innerwise.com, provides several videos and discussion on the technique. Your subconscious acts as a database collecting details about each and every memory of your past and even predictions of your future. To truly understand the concept of energy healing and why it is so important to health, you must recognize the body must be viewed from every perspective, even through space and time. This is called spherical vision.

Dr. Albrecht teaches that illness manifests from different things including not only physical distress but also emotional and other types of distress. Illness is affected, and sometimes even caused, by the energy of those around us. And, as my dog's homeopathic veterinarian states, energetic imbalances are the cause of much dis-ease. Additionally, illness is triggered at a particular time so in order to properly diagnose a problem we must be able to view the illness from many perspectives including through time.

Consider this concept when thinking about cancer, which will be discussed at length in a future book. We know we all have cancerous cells in our bodies. They are not foreign to our bodies but actually quite native. The reason we do not all "get cancer" is because of our immune systems and lifestyles. It has been said once cancer is visible or obvious, such as a lump in the breast, it has been growing for many years and thus your body was unable to do its job for some reason.

Brain tumors can begin many years before they are actually detected. This means the issue was triggered at a specific time and more and more research is showing that one can almost pinpoint exactly when it began simply by recalling the stresses one was under at that time. This is no different than the concepts Dr. Albrecht teaches. In order to truly clear out these energetic disturbances it is extremely useful to know when it began and therefore spherical diagnosis is important.

These ideas can be difficult to grasp at first. One of the examples he uses to highlight the importance of holistic healing, or viewing the individual as a whole, is imagining you are sitting down near a table with a bowl of fruit containing an apple. If you look at the apple from one direction it might look flawless because you do not see the wormhole on the other side. It is not always realistic or even possible to be able to see such flaws when physically "looking" at the patient even from various perspectives.

How can you know there is a worm and, more importantly, how can you know when the worm got there without spherical vision? If you do not know when the worm burrowed into the apple then you will not know how much damage it has done to the apple. The point is the problem cannot be seen by the naked eye and thus you must be able to look at it from various points of view, meaning not only through the eyes but also through intuition and with regards to space and time. There is no diagnostic tool within the mainstream medical system capable of doing this.

Using the method discovered by Dr. Albrecht once we learn how to trust our intuition and our own bodies then we are able to ask our bodies specific questions. Through our own muscle responses we may find answers to many things, including when we developed our illness, whether we are allergic to something that might be causing the illness, if there is some emotional connection with the illness with which we must first deal before finding true health, and so on. Using these tools one can also tap into his or her intuition to ask questions about making the right choices in everyday life. In the upcoming chapters you will learn how energy healing is rooted in science and some of the ways energies can interact.

Many naturopathic doctors and chiropractors firmly believe illnesses can be traced back to a single event or cluster of events. Even mainstream medicine is finally admitting how stress can cause illness. I believe it can be proven not only on an emotional and energetic level but even a physiological level and some studies are already showing how. Because our bodies have a "fight or flight" response to stress we can cause immediate injury to various processes within our bodies or parts of our bodies. For example, many people hold their breath while stressed. Either that or they have but shallow breathing without even realizing it. Doing such things deprives the body and brain of necessary oxygen.

Another major issue with regards to things like heart attacks is the clotting factor we each have. During trauma the body produces a greater number of platelets in anticipation of a traumatic injury. It is necessary to have this ability for the purpose of preventing excessive bleeding and hemorrhages during such injuries; however, having a high clotting mechanism all the time can cause a clot in already calcified or clogged arteries leading to stroke or heart attack. Since times of trauma, and therefore stressful fight or flight situations, can trigger the clotting factor it is critical to train the body not to be in fight or flight all the time so as to optimize the function of the clotting factor.

Unfortunately many people live their lives under constant stress and very minor issues can feel like a fight or flight situation. Such people have greater risk of illness such as heart attacks. These days this is not so uncommon. Sadly stress can trigger other dangerous disease without our realizing it until months or even years later.

In studying the metabolic theory of cancer, as proposed by Dr. Otto Warburg in 1924, we are presented with the idea that cancer cells begin as an injury to the cell due to a lack of oxygen. Once injury occurs the mitochondria becomes damaged and the respiration of the cell abnormal. Could the failure of deep breathing during times of stress contribute to the initiation of a cancer or tumor cell? There are no direct studies to say whether this is possible; however, if the metabolic theory of cancer is true, which is appearing to be the case as time goes on, we might be able to surmise this could be a cause in some people.

The body is complex and scientific studies can only provide so much knowledge of the body such that the knowledge can prevent or help cure illness. It would be much easier to be able to ask the body itself what is going on with it and get the answer directly from the innate wisdom of the body. This is absolutely possible with the energy healing tools developed by doctors like Uwe Albrecht.

Chapter 12

Energy Healing: Innerwise Part Two
Understanding Intuition

"Innerwise helps us understand the energy patterns and fields of people, animals, and systems that are at the core of life itself, and shows us ways to clarify and resolve irritations in the easiest possible manner."
-Uwe Albrecht, M.D.

In order to harness the "secret" power of energy in your life so you might make some positive change it is important to try to understand why and how it functions in our lives. It is one thing to say, "I'm going to do affirmations daily and envision my goals and one day I'll achieve them simply because I am doing this," but an entirely other to understand *why* this works. Achieving goals does not happen over night. Often it takes many weeks, months, or even years depending how big the goal is. Without understanding how energy works one might give up on using it believing it is not working. Rest assured your energy dictates your life much more than you realize.

Many people strive to live consciously and believe the conscious mind is responsible for their actions and lives. While "living consciously" is important it might have a different meaning than most people realize and they tend to forget about their deeper selves - their subconscious minds. It can be dangerous to forget we are not simply what we think. In fact, there has been much discussion and study on the id, ego, and superego. The ego is supposedly that part of the mind that cannot see beyond itself. In many ways it is like the conscious mind.

Living consciously is a good thing, as it can help people rationalize and come up with solutions to problems. It can also help people to seek a deeper meaning to life, but the question is, does that deeper meaning come from the conscious mind or from some even deeper place? If you study energy and healing it would appear so, as the deeper meaning can only come from the still small voice inside.

The conscious mind is rarely still. It analyzes and works through issues as though it is navigating through a maze. This could take much thought, rationalization, logic, analysis and therefore a great amount of activity. The subconscious mind operates differently and it is believed to be connected directly to the Source, or God. Moreover, as will be discussed in a later chapter, it is also connected to the earth, the sun, and to all other living beings. Through this connection the subconscious mind has access to all of the information from all of time; however, it cannot be understood simply from thinking. One must be still and listen.

To many, these ideas sound a bit kooky but rest assured they are actually well rooted in science. Many studies have been done on the subconscious mind and how it can store things we are not even aware of. It is the targeted location for subliminal messages, and the advertising industry is well aware of its power. If the most resourceful companies did not believe in this concept these subliminal messages would not exist. Billions of dollars would not be spent on an entire industry tapping into your subconscious mind if it were frivolous.

It is the subconscious from where intuition comes. When you get that "gut feeling" it is coming from your subconscious. They say this sixth sense, intuition, is stronger in woman than it is in men, which seems to be true, likely due to the more emotional nature of women. Although women always have a lot on their mind, they generally come from feelings rather than simply ideas or thoughts. Men tend to suppress their feelings opting instead to use their minds to find solutions to things. This is not to say men do not also have strong intuition. Children and babies likely have stronger intuition than adults because they have not yet been "corrupted", for lack of a better word, by the things of the world.

Moreover, humans are subjected to a great deal of "noise", which affects the function of our brain to the point where we struggle to alter the frequency of our brain waves enough to connect with ourselves on this deep level. This noise includes actual noise and material things of the world but, more importantly, it includes the electromagnetic pollution discussed in Chapter 3. Indeed native tribes, for example, were much more connected with themselves, the earth, and the animals than our generation could ever dream of being. It came naturally simply because they were not exposed to the technology we are today. This is why, unfortunately, we must work toward that connection, as it does not come naturally or easily.

Even animals, and maybe especially animals, have a strong intuition, and it may be stronger in animals than in humans, as this is how they function in their own lives. Dogs can sense if someone is a friend or a foe. If a dog has this ability, what would make anyone believe humans would not be equipped with it perhaps even to a higher degree, as higher functioning beings? With regards to health, animals are beyond wise in finding the remedy. Dogs regularly eat grass when dealing with an upset stomach. In the wild animals know which herbs are needed for various ailments. Clearly the animals are not learned in herbal medicine. This comes from their innate wisdom, the intelligence that can only come from the subconscious through intuition. As a matter of fact, ancient and not-so-ancient civilizations discovered remedies for illness by watching animals and following their lead with regards to medicinal plants. Many pharmaceuticals today are derived from some of those medicinal plants.

Unfortunately the establishment has suppressed knowledge of the sixth sense (and other senses) but there is no denying its existence. We have all experienced many moments of intuition which we ignored only to find out much later, sometimes years later, that our intuition had been correct.

How often have you made a decision after second and even third guessing yourself and you find out eventually that your first choice would have been the best? Our first choice is usually the best because it generally comes from our intuition, especially if we make the decision without putting much thought into it. When we allow our conscious mind to make the decision it will not always lead us in the right direction.

Women's intuition seems to be strongest when it comes to matters of the heart; like when their children are in danger or their spouse or significant other is cheating. I know of many women, myself included, who have ignored their intuition when they had a feeling a partner was cheating, but when the partner denied it they ignored the intuition and went on with life only to find out later their intuition had been correct and the spouse was lying. I say never ignore your intuition. It is your internal alarm system telling you something is not right.

Many a woman has nearly lost her mind simply because her cheating spouse convinced her conscious mind that her intuition was wrong. It is a rare gem of a man who can admit when he cheats. Most people will not admit to cheating unless he or she is desperate to get out of the relationship or when he or she is caught, otherwise he or she will deny to the bitter end. This denial is extremely damaging to the person who suspects cheating, usually a woman, even if her partner has admitted to, or she has caught him, cheating to a lesser extent than her gut is telling her.

Because her intuition is using her entire body to speak to her she knows deep within that either she is correct or she is crazy. When her logical mind is out of sync with her subconscious it can wreak havoc on her not only mentally but also physically.

Our bodies are always tending toward survival and that means they are always tending toward healing. It is my view that when people do not heal or end up succumbing to some illness it is due to their environment, nutrition or the fact that we are actively sabotaging ourselves by confusing our subconscious minds, such as with detrimental subliminal messages. Your subconscious mind seeks the best for you. It wants you to be successful and it will use whatever tools available to lead you there but ultimately if you allow confusing messages there will be challenges.

We experience the most suffering and hardship, emotionally and physically, when our conscious minds are not in tune or in sync with our subconscious minds. The body literally sends messages that your actions will not lead to success. This is true in all situations.

Imagine, for example, when a married couple considers counseling, or when one or the other seeks therapy after or while contemplating divorce. Perhaps the wife is the cheating spouse, which is something not in sync with her subconscious, as it knows this will ultimately lead to heartache. She is miserable in her life due to this or whatever led her to cheat in the first place. Perhaps she felt unfulfilled and unloved in the relationship and the more she sought that love the less she received it and another man was able to fill a void. Even prior to giving in to the other man she was not living the life her subconscious wanted for her. She was out of balance.

The husband is also out of balance because he too knows he is not living the life he is meant to live, as he should want to give his wife love and for both to be fulfilled. It could be as simple as a forced relationship from the onset or the couple allowed the so-called reality of the material world to get in the way by seeking more money, fulfilling the ego with things, and so on. Seeking or considering counseling is a sign that the conscious and subconscious are not in sync.

Consider the case of an illness. There are few illnesses we do not bring on ourselves by our unhealthy habits and even fewer we are unable to reverse by adopting healthy habits. Our medical system would like for us to believe that illnesses just happen and they can be fixed with a pill or surgery when, in truth, it was likely the lifestyle that led to the illness and it is the lifestyle that can make or break the person.

When it comes to chronic pain we have a choice of allowing it to consume us or use it as a tool to help us feel gratitude for the good things our bodies can do for us. Being miserable on a daily basis leads to further suffering because our subconscious mind does not want that for us. Pain does not have to equate to suffering. It is our body's way of getting our attention that something has to change, be it physically through diet or exercise, or mentally through positive affirmations, gratitude, etc.

In essence all things we feel within our body is our intuition speaking to us but instead of listening and trying to get to the bottom of it we mainly seek to mask it. Sometimes the feelings are "good" like butterflies when meeting a new potential love interest. Still, that feeling is your subconscious telling you that you should pursue this relationship. If you do not then your body is going to speak to you in other ways, perhaps with sadness or regret.

Intuition manifests in different ways but the bottom line is that our bodies are capable of communicating with us not only in terms of premonition but even to confirm feelings of which we are already aware. We experience goose bumps, sweaty palms, increased heart rate, "butterflies" in the stomach as mentioned, and other things in response to certain situations. These reactions serve to cleanse our bodies of certain energies and to communicate with us to help us decipher situations. This is important because it keeps us safe and leads us to success, as long as we are paying attention and as long as we do as our bodies say. Of course our bodies aren't going to speak in words therefore it is important for us to try to understand what these reactions mean so we might figure out the best way to deal with issues for the best possible outcome.

When we make a decision that is not in sync with the subconscious mind the body feels it and will become imbalanced. This can happen even if we do not actually make the decision but just consider it. Most of the time we do not know the body is imbalanced or that we are making the wrong decision because it can be subtle; however sometimes we feel something very distinct. Have you ever done something you know is wrong and when someone calls you out on it you feel a distinct sensation in your body? That is your body's way of saying it is imbalanced because your conscious mind did something not in tune with the subconscious. It cannot be stressed enough that your subconscious knows what is best for you.

These forms of communication are nothing more than the energy fields within your body acting either in harmony or disharmony. When the energy fields in our bodies are in disharmony, and when our conscious minds are not in tune with our subconscious minds, our muscles also sense the imbalance.

Through Chinese medicine we know our bodies are separated into two meridians. Each meridian is connected within itself but the two are not connected with each other. Also, scientifically we know our body is always tending toward balance, which is called homeostasis. This is true for all functions of the body, whether it is our kidneys balancing the load of filtration between the two, our cells ensuring the proper balance of electrolytes within and without the cell, or our blood seeking a neutral acidity and the proper specific gravity. Of course in order for our bodies to achieve perfect balance we must be healthy. Being healthy means our conscious mind is in sync with our subconscious mind, which manifests in all ways, physically, emotionally, mentally, etc. and one way to know this for sure is if the muscle responses between the two meridians are in balance.

If there is some internal resistance to a temptation, the muscles also become weak. It is possible to tap into your intuition and train your body to communicate with you directly through these muscle reactions to receive direct answers to specific questions. This takes some practice but it is possible and many people have mastered the art. Read on for a discussion on some of the early tools used for just this function and how Dr. Albrecht began using the body itself as a tool and is now teaching others to do the same.

Chapter 13

Energy Healing: Innerwise Part Three
How To Tap Into Your Intuition

"The Arm-Length Test, Yes or no: Life can be
that simple! This isn't an esoteric fad, but rather,
a neurological reflex, the body's own biofeedback
system, with the muscles, tendons and fasciae
relaxing on one side of the body while
contracting on the other side."
-Uwe Albrecht, M.D.

Although your subconscious mind is never wrong when
it comes to knowing what ails you, when it began, or how
to address it, there can be times when it is unable to
properly communicate its message through intuition. This
happens when the body is under stress, which can confuse
the lines of communication from your subconscious mind
through your intuition. Fortunately there are many
techniques that can help get the body back in sync.

Practitioners who use the body's innate wisdom in their
practices are able to treat the body as a whole quickly and
efficiently. For example, Dr. Storkan is able to treat various

conditions provided they are not too far advanced; however, even if they are advanced and require more aggressive treatment he is still able to determine the best course of treatment because it is the patient's own subconscious that is directing the doctor.

In terms of ensuring he is receiving the correct message Dr. Storkan begins by asking his patient to place one hand on top of his head with the palm faced down while he pushes down on the opposite arm. He then asks the patient to reverse his hand – palm facing up – while pushing down on the opposite arm. During this test he "asks" the body questions to determine whether or not his patient's subconscious mind is in sync with the conscious mind meaning, whether or not he can trust the body's answers.

The results of this test can indicate whether the body's responses might be inaccurate due to stress. If the arm is able to resist the push of his hand while the opposite hand is placed palm down on top of the head then the body will provide correct answers and the nerve impulses are not "misfiring", so to speak. If the opposite happens, however, there are imbalances and emotional blockages must be cleared. This misfire occurs when there is stress. Usually it indicates that the patient has been dealing with actual stressful situations, which can result when the subconscious mind has been receiving confusing messages. The stress occurs, for example, because the conscious mind seeks to overcome the subconscious with its own agenda.

It also could indicate that the stress has been so great the individual "checks out". This is one of the dangers of failing to connect with oneself. When we try to avoid and are unaccustomed to facing our own fears, worries, failures, or shortcomings it can cause this type of effect. We think running away from our feelings will make things better but it can actually make things worse and cause pain, illness, and further emotional turmoil.

Dr. Storkan employs a form of muscle testing called kinesiology, which he uses for all diagnosis. The imbalance in the body is realized when he places force against a limb of the body such as the arm or leg. Through Innerwise, however, each individual learns how to do this himself using his own body and therefore could potentially treat his own ailments as well as using this simple but powerful technique to direct himself along the best and most successful course.

With regards to the arm length test used with the Innerwise program this muscular imbalance presents itself with a difference in the length of the arms. In essence, what these diagnosis modalities are doing is tapping into our own intuition, the sixth sense we discussed, which then allows us to communicate with our subconscious minds. Again, to many this might sound outrageous and difficult to believe but can be proven through science as previously discussed.

Because of the nature of intuition we cannot "see" it nor can we usually verify what it tells us, at least not immediately. The danger in not following our intuition or "gut feeling" is that we are usually squandering an opportunity but, as mentioned previously, it could be as detrimental as full-blown illness and mental anguish. One should always trust her gut when it speaks to her. It will never do her wrong. The first reaction is always the correct reaction because it is the intuitive reaction.

However, keep in mind that the messages sometimes get crossed when it comes to our subconscious minds speaking to us. The only way to really know if this is happening is to pay attention. People will say things like, "I was sure this would work, where did I go wrong?" What likely went wrong is that the lines of communication became crossed due to stress within the body, which can be corrected very simply, for example, by using the Emotional Freedom Technique or EFT, a topic that will be discussed in more detail later.

This technique requires repetition of a type of positive affirmation while tapping on a certain part of the body. In Dr. Storkan's practice he has the patient use three of his fingers on his right hand - the index, middle, and ring fingers - to tap on the outside of the left hand while making the statement, "even though I feel miserable I deeply and completely love and accept myself," which must be said three times during tapping. This method has actually been found highly effective in any stressful situation in which an individual finds himself. EFT works by tapping along meridians identified in Chinese medicine and is designed to allow your body to accept healing much more quickly than it would otherwise. It essentially embeds your subconscious mind with the message that you are deserving of all healing and thus directs the subconscious to seek the positive, successful route toward healing.

The method taught through Innerwise, is to continually use the arm length test while stating, "I am I" until both arms are the same length. In some cases a person might be going through very stressful times and thus might require to do this exercise regularly, even daily for weeks, until he is sure the messages sent through his nerves are in sync with his subconscious and thus the "wires" are not crossed.

The purpose of this particular discussion is simply to explain that there is such a thing as having a type of polarity reversal with regards to the messages sent from the subconscious and thus if we are suffering from a great deal of stress we might not be able to trust our own intuition if not addressed.

It is important, however, to first realize the damage caused by stress in this, and other ways, and to know there are highly effective ways to deal with it. Recognizing when we are under stress and actively working toward dealing with that stress is critical.

Of course most people do not understand that the feelings in their bodies, whether physical or emotional, are the messages being sent by their intuition. Unfortunately because of the lessons ingrained in us from birth to ignore and, especially, not seek understanding of the energetic aspect to ourselves they fail to realize this would greatly help them to heal if only they would stop masking symptoms with drugs. This, alone, is a message I hope to convey with this book but for those seeking even deeper understanding and optimal success I am here to say there is a way to use your own subconscious not only to heal from whatever ails you but also to find ultimate success and happiness.

So now the question becomes, how can I read my body's intuition in situations that are not "fight or flight"? Our bodies are very good at sending us the message of danger or discomfort, the struggle is in having the ability to tap into our intuition at any given time. The good news is there have been discoveries of various tools, instruments, muscle reflexes and energy fields, which can help us see and verify our intuition.

Such tools have included: 1. A Pendulum - the direction in which the energy field of the palm rotates changes with a 'yes' or 'no' statement. 2. Dowsing Rods - these can be made of wood or metal and also react to the energy field of the palm. 3. Biotensor - also reacting to the user's palm, these instruments are very sensitive, which is amplified by using coiled wire inside the handle. 4. The Lecher Antenna – a hollow antenna rod developed by Austrian Physicist Ernst Lecher, which can be adjusted to set the antenna to specific frequencies. 5. The Muscle Test – the kinesiological test I have been discussing as used by Dr. Storkan, which uses the muscles' own weakness to resistance as a measure; a negative statement renders a strong muscle weak. There are also other tests using the human body as the tool including the Pulse Test used mainly by acupuncturists, the O-Ring Test, and the Standing or Sway Test.

The way Dr. Albrecht came up with the Arm-Length Test is fascinating. I highly recommend reading his book, *Yes/No, Using the Arm-Length Test for Instant Answers and Wellbeing* to learn not only about the test but also about how it came to him. In fact it was originally discovered accidentally by a Belgian osteopath named Raphael van Assche when working with a patient.

We are all aware on some level that our bodies "know." They just do. We are aware of the self-healing capabilities of our bodies thus it stands to reason our bodies know much more than we think. We have all experienced the "gut feeling" and our sixth sense from time to time. It is not practical to see a doctor who practices kinesiology whenever we need to make an important decision. It would be ideal if we were able to get those answers immediately, as soon as a question arises. In the next chapter you will learn how you can use your own body as that tool.

Chapter 14

Energy Healing: Innerwise Part Four Combining Healing Tools, Healing Frequencies and the Power of Intuition

As previously discussed, various types of tests help diagnose problems and illness by tapping into the subconscious, which is known to hold the mind and body's "secrets" so to speak. The subconscious is where essentially all information is stored including memories, emotional trauma, and information acquired from external sources like television, books, etc. For this reason it is incredibly important to be cautious of what you allow into your mind.

Television is well known for inputting certain types of messages into the viewers' minds. There is a reason it is called "programming." Advertisers and producers alike have learned how to tap into various areas of the mind to allow certain ideas to be fully ingrained without you really knowing they exist. Have you ever had a sudden craving for a Big Mac and you have no idea why? It is entirely possible that your subconscious held the message of a McDonalds commercial, which led to your craving.

I previously discussed the dangers in having a television in the bedroom. In addition to the physiological implications, as discussed, having a TV in your bedroom makes it very likely that you might fall asleep while watching and, more importantly, listening. Although there are not enough studies to corroborate a clear connection between sleep and subliminal messaging there are anecdotal stories indicating people in an unconscious state are absorbing the noises around them.

For example there have been stories of comatose patients being prayed over resulting in complete recovery due to the patient realizing he or she had a purpose for living. The patient could pinpoint when this desire to live came about, which always correlated with the timing of the prayer.

I have a friend who told me the story of her father who had a heart attack at an already elderly age. He ended up in a coma for several days. The doctors gave a grim prognosis and did not expect for him to survive, especially considering his age of being over eighty-years-old.

During one family visit while he was in intensive care at the hospital the nurse informed them that the gentleman in the next room who happened to be Native American had passed. His family would be coming in to have a ceremony with singing and drums and the nurse did not want to disturb my friend's family so she informed them she would close the door. Even so the music came through the walls.

The next day my friend's father suddenly woke up. "I'm hungry! I want a cheeseburger," he proclaimed. The family was overjoyed and beyond surprised and when he was asked what happened he said, "Well, I heard this beautiful music of drums and singing and it made me think about how nice it is to be alive and I decided I wanted to wake up so I could enjoy more music. I realized I wasn't ready to die." The family realized he had heard the music coming from next door.

There have been many stories similar to this in which coma patients have awakened due to great stimulation from prayer or music. As it is we know our brains are very susceptible to subliminal messaging while we are awake, which can prompt us to take certain action such as giving in to the Big Mac craving, however, anecdotal stories as well as some studies have confirmed that messages making their way into the mind during ultimate unconsciousness such as a coma can prompt complete recovery if the patient is given positive messages.

When we are awake the brain exhibits beta waves, those ranging between 12.5 and 30 Hz. In a state of relaxation or meditation our brains exhibit alpha waves, between 4.1 and 12.4 Hz and in deep sleep the brain exhibits delta waves, 0 - 4 Hz. Studies have shown that during television watching our brains drift into alpha waves and even close to delta waves. It is bad enough that our brains are absorbing messages we might not normally choose while viewing television. Imagine the messaging being sent into your mind while you sleep in the company of a television streaming who-knows-what.

Indeed we are exposed to much more information through the TV than we ever would be in real life. We become aware of concepts we otherwise would have never dreamed of and while some of these concepts might be educational some are quite dastardly. Your subconscious mind stores all of this as memories whether or not you lived them.

There is a separate, parallel, but similar reasoning explained with regards to the existence of negative entities such as those referred to as fallen angels by people of certain religions. A future sequel to this book will include a discussion on this concept. To highlight this briefly, however, it would be beneficial to share the idea that exorcist priests are well aware of the way these entities operate and the things they are "allowed" to do. For one thing these negative or "evil" spirits are allowed access to all of your memories and to use them against you. Of course if the entity is a negative or sinister energy it would stand to reason it would seek to damage you with a negative memory. Does this not sound like the scientific explanation of subliminal messaging and the subconscious mind?

It is my belief that all things energetic can be explained *both* scientifically *and* spiritually and neither one is more correct than the other. In researching this topic it is obvious that these phenomena are both scientific and spiritual, two concepts that cannot be divorced from one another. For purposes of this book, however, I choose not to go into detail on the negative aspect of these entities because even just thinking about them gives them additional energy. In an attempt to demystify these concepts the message I prefer to convey is one of a positive nature such that my readers are able to use this information to benefit them greatly including physically, emotionally, and mentally.

Expanding on the discussion above, as the database for all memories, the subconscious is the part of the mind that knows the exact cause of an illness, whether it is due to an external toxin, an emotional trauma or a physical injury or, quite possibly, all three. For this reason it is highly beneficial to learn how to ask your subconscious the right questions, but it is necessary to be able to speak its language; the language of energy in the form of imbalances. This is your intuition.

The arm length test is one of the best ways to do this because it does not necessarily require a second person yet it allows you to ask very specific questions and to practice as often as you'd like. Unlike the presence of sweaty palms, butterflies in the stomach, and an elevated heart rate, which are present in undeniable situations of fight or flight the arm length test can detect subtle changes in the body due to unknown and unseen forces. The most difficult aspect of learning to tap into your intuition is learning to trust yourself; however, with practice it becomes easier.

With the Innerwise system, which is considered a medical treatment, the same arm length test used to diagnose the problem is also used to determine the proper treatment. Along with "healing" cards, treatment is done using an amulet. As with the cards, it could be concerning whether or not this amulet is considered a tool for divination, which is forbidden by certain religions. What is clear when you look more closely is that the entire process is based on science and the very real, albeit invisible, concept of energy.

In studying other healing modalities, such as homeopathy, it is also clear that the power of healing that comes from the energies of various essences is also present in Innerwise although they are infused in different ways. The energetic frequencies used in homeopathy result from the dilution of harmful substances, those that cause the same symptoms as certain illnesses. The cards used with Innerwise have been embedded with the energetic frequency codes of various plant essences, colors, animals, celestial bodies, etc. including some of those used in homeopathy.

It is believed that the energetic frequencies from the substances used to make homeopathic medicines remain in the memory of the water used to dilute the substances. It is this remaining energetic footprint left in the water that is thought to be the reason for the efficacy of homeopathy. By all accounts it is the energetic frequencies that cause a minor irritation triggering the immune response to deal with those symptoms. The same holds true for healing cards.

The arm length test is used to determine the number of healing cards required to treat the issue then the practitioner's or individual's intuition is used to determine which particular cards are necessary; the first choice is always the correct choice. These cards are then placed upon the amulet as the frequencies are transferred using the energies in the palm. The amulet is then worn next to the body so the individual absorbs the healing frequencies. Sound crazy? Consider that energy cannot be seen but it is there all the same. The arm length test is then again used to determine how long the individual must wear the amulet. More seasoned practitioners are able to make these determinations on the first try while others literally need to practice to have repeatable results.

The arm length test has been effectively used to determine such specific things as glucose levels in diabetics and blood pressure levels. For practitioners, using the test to determine verifiable results is useful to ensure accuracy in other areas. Innerwise has been effective in treating a whole range of illnesses, even kidney disease to the point of complete resolution. Energy imbalances can have not only a great effect on health but also on emotional well being, which can then lead to health or sickness.

Chapter 15

Energy Healing: Innerwise Part Five Understanding Energies And Using Intuition as a Protective Measure

Although this topic can seem a little off the wall the fact is that we are all very aware on a daily basis of the invisible phenomenon we call energy. We say things all the time relating to energy in our experiences such as, "I have no more energy", "He/she sucked the energy right out of me," "He/she had bad energy", "I love your energy," or "I have so much energy today".

When you learn about energy you will find it interesting how many situations and experiences become clear. Indeed external energies, such as those called upon in practices such as divination, whether good or bad, do exist. Those who believe in Christianity would call those energies either angels or demons. Dr. Albrecht simply refers to them as positive or negative energies. He clearly explains how Innerwise deals only with individuals and their energies and cautions against comingling energies even between two people let alone welcoming the supernatural energies.

When a practitioner works with another individual she must be careful to "disconnect" from that other person's energy at the end of the session so as not to carry his or her energy with her.

In fact we are constantly comingling our energies with other individuals without even realizing it, which can cause a great deal of turmoil. This turmoil can result in excessive stress leading to illness and an overall poor quality life. In actuality we are all better off protecting our own energies from others no matter who those "others" are.

Dr. Albrecht used several examples and stories of how this all works. There are high energy and low energy people and there are also people with high charges and those with low charges. These are separate concepts. High and low energy means just that: a person with high energy has a lot of stamina, higher thinking, highly productive, etc. while people with low energy seem tired all the time. This could be measured, determined, or even just discussed in terms of frequency. The higher a person's frequency the higher their consciousness and the more they are able to accomplish and solve. Remember, however, operating at a high frequency all the time is not desirable as healing occurs in lower frequency ranges, such as the delta waves of deep REM sleep (0 – 4 Hz).

High and low charge means something different, much like molecules or batteries. Highly charged individuals are constantly looking for those with low charges in order to "discharge". They need to blow off steam; get rid of some of that charge, much like free radicals seeking an electron to close its charge so there is balance. Again, the body is always tending toward homeostasis.

Consider when two people are in an argument. Their charges are constantly being transferred from one to the other. Next time you find yourself in a dispute with someone, remember this and do not allow the individual to discharge on you. If you are able to maintain your low charge, the charge of the other individual will just eventually dissipate into the air without escalating the issue and you will both soon have a low charge. Those with high charge are always looking for someone with a low charge upon whom to discharge. It is up to you to be aware of this and not allow it to happen.

With regards to energy, on the other hand, we want to have a great deal of energy and to keep our energy with us and not allow someone else to "suck" that energy up. Certain situations can be emotionally draining and can take our energy from us. After dealing with certain types of people we actually feel drained of our energy. These terms are not accidental, nor is it accidental for us to physically feel the effects of energy, positive or negative, after interacting with certain individuals. In fact certain people seek out others from whom to steal energy. These people should either be avoided at all costs, or you must learn how to deal with energies so as not to be harmed.

Practicing to tap into your intuition using tools such as the arm length test allows you to hone this skill to the point where you are able to sense energies without even having to use the test. Of course this takes a great deal of practice, but it can be done. Some people can sense the energy state of others simply by speaking to them and sometimes just thinking about them.

Dr. Albrecht gives the example of walking into a party with many people and mingling with various individuals. You may have entered the party with a great deal of energy and by the time you are ready to leave you feel sucked dry. It is likely that you "left" some of your energy with various people - those who prey on others' energy, whether knowingly or unknowingly.

By learning how to effectively use your intuition you can learn to actually see your own energy by assigning it a specific color. Using this technique you could scan the room just before you leave the party and actually *see* where you "left" your energy and thus reclaim it to be made whole again. This might sound crazy but it actually works. Based on what Dr. Albrecht discusses it seems this could be what those who can read auras have learned.

Looking back to the chapter on the dangers of cell phone use and how to protect not only your sleep space but your body in general, it would seem having the ability to connect with yourself to the extent of being able to "see" your own energy would be highly beneficial in even protecting your cells from damage caused by rogue signals.

In addition to using these concepts other energetic healing modalities have proven highly effective in treating illness as well as effecting overall wellbeing simply by setting intentions (think Law of Attraction) as well as envisioning healing through energetic means such as colored flames, e.g. the violet flame or other colors of light. Practices such as Reiki also utilize intention as a source of energy for healing.

Chapter 16

Energy Healing: Innerwise® and Intuition Some Final Thoughts

As you can see, there is much to discuss about energy healing. Although there are other modalities and programs that help understand energy, intuition, and healing, Innerwise has been one I have understood the best because it is presented in scientific terms and can be correlated with scientific concepts while being simple to practice with or without a coach.

After having read this you are by no means an expert, however you are far ahead of the curve in understanding the importance of protecting the subconscious mind, learning how to tap into your intuition, and why energetic healing is key to overall health and success.

One important concept to take from these Innerwise chapters is the scientific and medical basis behind energy healing. It is a very real issue contributing to our health problems today. It is no surprise that, as society has become more negative and even weak in the mind, it has also become less successful and more ill. According to a Business Insider headline from February of 2016,

"Something startling is going on with antidepressant use around the world." The mental health of an individual affects the mental health of the society and energy plays a huge role in this aspect.

Energy is always commingled with the collective therefore it is more important than ever to protect your own energy and ensure you are surrounded by people with a positive attitude. Also, your positive (or negative) energy and attitude can affect everyone.

Our intuition, with its basis in the subconscious, is intimately connected with our own energy as well as that of others. Intuition not only influences our relationships with others but can influence entire groups of people and even countries, for example by affecting public policy. In fact there are some Innerwise practitioners who use the system with corporations and businesses. It could be considered a sort of "feng shui" for organizations to ensure proper and healthy energy flow for optimal success. This can be extremely useful when determining hiring or even placement of potential employees.

The usefulness of this tool regarding systems and organizations makes perfect sense considering the results of comingling individual energies with the energies of others. Obviously being energetically in sync with those around us would make for an ideal work environment. Imagine the immense possibilities if the majority were "living the abundance of their being." We could potentially envision even a utopian society. Conversely imagine the ramifications of the majority heading down a slippery slope of misery, self-medicating with pharmaceuticals, drugs, and alcohol. Perhaps it is the effects of this state of mental health we see today.

Closing out the chapters on Innerwise, here are some final thoughts about energy healing, Innerwise, and intuition.

First, on the topic of commingling energies, one of Dr. Albrecht's most interesting discussions was regarding allowing negative energy into an individual's space. He did not discount the existence of positive or negative energies; what many of us might call angels or fallen angels. It is always interesting when those from the scientific community acknowledge this type of phenomenon, essentially acknowledging the spiritual world.

In discussing the example of entering a room full of people, such as at a party, where you find you are either much more energized or absolutely drained after having a conversation with a particular person Dr. Albrecht goes on to discuss these external, negative or positive, energies. One thing he discusses is how those who drain you of energy, particularly when they speak in negative tones or words, allow the opening of portals to other dimensions. These portals allow negative external energies to enter into our world and wreak havoc.

Many people find this to be superstitious or nonsensical but, as mentioned before, the inability to see energy with our eyes does not mean it does not exist. As such, the inability to see portals with our eyes also does not mean they do not exist. A discussion on the varying frequencies, and thus the various potential realms and planes of existence is detailed further in a later chapter.

Because we cannot see frequencies or energies it is extremely important not only to be mindful of the words we use, tone of our voice, and our thoughts but also we must be careful who we allow into our space, his or her demeanor, the words he or she uses, and his or her tone of voice.

Having a conversation with someone who is negative, spewing hateful things, talking negatively about others, etc. not only zaps our energy but it can also open those portals allowing negative entities to enter our space and even our bodies. The term "possession" means just that - a spirit in possession of another's body, a phenomenon that has been witnessed by many highly educated theologians, such as exorcist priests, and can actually be explained in scientific terms.

I have had discussions in the past with Christians who firmly believe illnesses such as cancer are actually evil spirits possessing the body. Although Dr. Albrecht does not describe it in these terms, he seems to agree. Various illnesses can be caused by emotional trauma manifesting as physical symptoms including heart problems, liver and kidney issues. Harboring negative thoughts and emotions, speaking negative words, holding on to grudges and anger regularly causes illness. It is arguably likely that illnesses are not just cellular changes in the body caused by stress or our own minds sending certain messages and signals but also a result of negative external energies we allow into our bodies via the portals we open through our own negative actions.

For more on the spiritual aspect of the subject conduct a YouTube search on Father Chad Ripperger, Ph.D. a long-time Catholic exorcist priest who speaks on the topic regularly and is one of the most knowledgeable in the area of angels and demons. Keep an open mind. Remember that just because Catholic priests happen to be part of a religious or spiritual organization does not mean they are not highly educated, even in science. Few Christian pastors are even educated, let alone as highly educated as Catholic priests therefore, in some ways, the knowledge of these learned theologians is more credible. This is certainly something to consider.

The good news is that using intuition to diagnose, an Innerwise practitioner can determine what type of trauma caused the illness even pinpointing the day it was caused. This is the "spherical vision" discussed through Innerwise. The energies causing the illness are likely lodged in the body; however, once these energies are cleared and the natural energy exclusively belonging to the body begins to flow freely without obstruction, true healing can occur including healing of the body, mind, and emotional well being. Moreover, once we realize that our subconscious mind, is connected directly with our Source and with all living things we can use it to lead us toward great success, which can not only help in our own personal lives but society as a whole.

Thanks to a YouTube video I recently stumbled upon – it seems this topic is gaining in popularity; either that or the Universe (God) is directing me to even more information supporting this topic – I have been made aware that many successful individuals attribute much of their success to intuition and their subconscious mind.

In fact Nikola Tesla, who I briefly mentioned before as being a highly intelligent inventor way ahead of his time made many statements regarding his connection with his subconscious mind and how it led him to his inventions and ideas.

"My method is different. I do not rush into actual work. When I get a new idea, I start at once building it up in my imagination, and make improvements and operate the device in my mind. When I have gone so far as to embody everything in my invention, every possible improvement I can think of, and when I see no fault anywhere, I put into concrete form the final product of my brain."

"My brain is only a receiver, in the Universe there is a core from which we obtain knowledge, strength and inspiration. I have not penetrated into the secrets of this core, but I know it exists."

"Every living being is an engine geared to the wheelwork of the universe. Though seemingly affected only by its immediate surrounding, the sphere of external influence extends to infinite distance."

"IF YOU WANT TO FIND THE SECRETS OF THE UNIVERSE, THINK IN TERMS OF ENERGY, FREQUENCY AND VIBRATION"

Ralph Waldo Emerson, Aristotle, Steve Jobs, and even Oprah Winfrey were well aware of the power of intuition.

"Trust your instinct to the end, though you can render no reason." – Ralph Waldo Emerson.

"Trust your instincts, intuition doesn't lie." – Oprah Winfrey.

"Intuition is the source of scientific knowledge." – Aristotle.

Steve Jobs had much to say about intuition. "Have the courage to follow your heart and intuition. They somehow already know what you truly want to become. Everything else is secondary." "Much of what I stumbled into by following my curiosity and intuition turned out to be priceless later on." "Intuition is a very powerful thing, more powerful than intellect, in my opinion." And, finally, "I would trade all of my technology for an afternoon with

Socrates." – Steve Jobs. Aristotle was a student of Plato and Plato was a student of Socrates. Perhaps Steve Jobs wanted to discuss these concepts with the man behind the man behind the man who believed that intuition is the source of scientific knowledge.

Although Albert Einstein did not seem to directly credit his knowledge or inventions to following his intuition, he certainly understood the power in energy. "Everything is Energy and that is all there is to it. Match the frequency to the reality you want and you cannot help but achieve that reality. It can be no other way. This is not philosophy. This is physics." Clearly he was discussing the fact that our reality is due to the vibration and frequency of the things we can see being in tune with our minds; however, as he states, we are able to create the reality we want by tuning into those frequencies on purpose.

Chapter 17

Collective Energy
And Its Effect On A Community

Old Sundevil Mascot   New Mascot

At this point it seems a discussion on a somewhat controversial topic, which made its way to my community, is necessary. The rural area in which I live is lovingly referred to by a local media personality as "the beautiful Española Valley" both for its aesthetic beauty as well as the genuine beauty of the people. Unfortunately the area is also very depressed and oppressed, with the majority of people living at or below the poverty level.

As in many rural communities these days, Española and its surrounding communities have been in the midst of a major battle with drug use. As time goes on and as generations pass the difficulties facing the people seem to become more and more dire. As a whole, the community has been in decline over the past several decades and, despite the altruistic efforts of a few well-meaning individuals, it does not seem to have the ability to pull itself out of the rut into which it burrows deeper and deeper.

If you are open to it and reflect on situations, you can find similarities within groups of people, organizations, communities, everywhere. The goal of this discussion is not to be controversial but to insert a point of view that many people likely have never considered when it comes to such things as the collective energy of a community and how it might be affected by something like a high school mascot.

Energy is a complex phenomenon affecting many things. It has been discussed how our health and wellbeing are greatly affected by energy and, in turn, how energy is affected by our thoughts, words, and actions. Our minds are very powerful. Science proves our minds have the ability to alter healing by sending certain impulses to our nerves. Our moods and sense of being affect these impulses.

When I first heard about the proposal to change the mascot of the local high school I immediately thought about the energetic aspect. Although the Española Valley High School, EVHS, was my own alma mater and I have always been a great supporter of the "Sundevils" I found myself in favor of changing the mascot after having studied energy and its incredible power over all aspects of our lives.

I once was one of those diehard fans chanting "devils devils devils" at games until I learned about the concept of energy. I know many people believe it silly to change such a longstanding mascot, which I myself will admit has a pretty cool logo, appropriately named "Sparky". But hear me out.

I was pleased a mascot change was being considered but a bit disappointed by the lack of explanation as to why. I had several Facebook friends who were weighing in on the subject mainly on the side of keeping the mascot claiming it would be ridiculous to change it and the school board should be worried about "more important issues". While these arguments are valid and made much sense, nobody was talking about the energetic aspect.

This is where I say to anyone who has no idea what I am talking about: an understanding of the contents within this book is necessary. I honestly believe you will get it once you have that understanding. Here's the thing, devils are negative energies, both in the "religious" or spiritual world as well as in the scientific world, as I discussed regarding the Innerwise® system. Most people in this valley believe in the spiritual world and many are very "religious." I, personally, am not speaking from a place of religion but from a place of logic because spirituality and energy in the scientific sense are one and the same. Still, it is not necessary to believe in religion at all to understand what I am attempting to convey.

If words have the ability to affect energy, which can affect wellbeing, then it stands to reason negative words cause negative energy, which in turn negatively affects wellbeing. Collective energy from a large population pervasively affects the wellbeing of an entire community. Many have written on this subject at length; it was touched on before and it will be discussed further in a later chapter.

With this being said, it has been some time since I have believed the wellbeing of the community of Española including its image, the drug use, and even the reputation of the high school itself and the success – or lack thereof – of its students is directly correlated with the negative connotations associated with the "devil" mascot or, at least affected by it. When the school was originally created the mascot was a fairly scary man-creature. The mascot has been seriously downplayed since the school opened in the late '70s, thank goodness, when it was such a sinister character that, as a child, I was afraid to look at its image on the cover of my siblings' yearbooks.

Regardless of the image itself, however, chanting and thus invoking a negative energy - a demon in fact - is absolutely affecting the community as a whole, regardless of whether or not you are religious. The negative connotation itself is enough to affect the collective energy of the community.

In a spiritual sense, it is well known in the Christian and especially Catholic community, particularly among exorcist priests, that speaking the name of an evil entity gives it more power, i.e. energy. According to Fr. Ripperger these entities do in fact have names and invoking or simply just speaking the name empowers it to do whatever damage it is allowed to do, or that it seeks to do. Therefore a collective chanting of the name or even just reference to the type of entity would most certainly give it energy in whatever space it is spoken.

The community has long been known for its support of the high school's basketball program. For as long as I can remember, the varsity teams, both boys and girls, pack in the crowds. During these games people are chanting to the negative energy over and over again. A chant is intended to glorify and, in this case it is easy to see who is being glorified. In the meantime the school itself has a graduation rate that has been declining for decades and is now at 59%. Coincidence? Possibly, but I highly doubt it.

It is much like the discussion in the previous chapter, when a group of people, whether it is a family, community, or society whose thoughts are primarily of a positive nature it is visible in the health of that system. When the thoughts and attitudes of a group of people, including a community rife with drug problems and poverty, veer toward the expectation of more of the same they receive more of the same. I couldn't say how many times I have heard people confirm and even justify the drawbacks of Española. I have been guilty of it myself. I firmly believe if we were to reject the notion of, "this is just how it is" and start changing our mindset and attitudes we could effect great change.

Change is difficult, but perhaps a makeover of the mascot could help facilitate the positive by turning the collective energy around simply by encouraging the community to speak the name of a positive entity or concept, which could effect positive change for the entire community. I think it's worth a try.

Chapter 18

Energy and Water

The discussion of energy as it relates to water is appropriate, as it will highlight the effects of the energy from words and music on the quality of the water in our daily lives, particularly the water existing within our bodies and thus its effect in our own lives, not just health.

The late author and holistic doctor from Japan named Masaru Emoto wrote several books on water crystal healing. Dr. Emoto's work centers on crystallization of water under certain conditions and in certain environments to determine the effects of words and music on the quality of the water.

To determine the effects our thoughts, words, and even the music we listen to, have on our bodies, Dr. Emoto conducted several experiments in which he placed certain words and phrases against petri dishes of water and allowed the water to freeze, or crystallize, while exposed to such words and phrases. In other cases he allowed the water to freeze while being exposed to certain types of music, namely classical and heavy metal, and also while people made certain statements and spoke certain phrases.

The water exposed to beautiful words such as "love", "gratitude", and "forgiveness" crystalized into beautiful and symmetrical shapes while the water from the same source but exposed to ugly words such as "hatred", "you fool", "you make me sick", and other similar statements crystalized into distorted and unpleasant shapes. He also determined that the way a statement is phrased could affect the form of the crystal. For example, "let's do it" indicating solidarity created a pretty shape, while the command, "do it" created a distorted crystal.

When it came to speaking the words or exposing water to music, the longer the water was exposed to a certain type of phrase or music, the more or less beautiful it would become. A crystal formed while children spoke the words, "You're beautiful" several times versus that formed with only a few repetitions of the same words was much more symmetric and aesthetically beautiful. Consider this when thinking of affirmations.

In most cases the crystals formed when the water was exposed to classical music were symmetric and beautiful while those exposed to other types of music, namely rock and heavy metal were typically distorted and ugly. Given the discussion in this book it is not surprising that ugly words would do such a thing considering the downward spiral negative words and energy have on every living thing. Most people have heard that even plants grow more beautifully when spoken to in kind words. It seems this research by Dr. Emoto confirms the power our words and even music have over our health and psyche.

The reason this type of work is so critical is obviously because we as humans are made up primarily of water and water affects our lives in immeasurable ways. A fetus is made up of approximately ninety percent water while the average adult body is made up of seventy percent. If the words and music to which we are exposed can physically change the structure of water in the solid state so completely, to the point where it actually becomes physically distorted, then imagine what those words and music are doing to our bodies on a cellular level. Moreover, the topic of energy in water is exemplified by homeopathy and the theory of how and why it is so effective; namely, the way water has the ability to store energetic information about everything with which it has ever come into contact.

I have thought about this concept often when observing the packaging of certain types of drinks including soft drinks and energy drinks, as well as wine, beer and spirits. Consider the "Monster" energy drink. It is said that the label shows the Hebrew letter "vav" or V/W, which signifies the number "6" and it is repeated three times thus giving the name of the "Beast", as discussed by those with religious beliefs. It sounds a bit like a conspiracy theory and it very well might be but either way it has a negative connotation therefore in terms of the present discussion we must ask, is it worth ingesting something that has been 'infused' with negative energy?

As craft beer and boutique wine has become more popular, so has developing more and more clever names and more and more artistic labels. A friend who brews his own craft beer gave his brewery an interesting name having to do with bones, or skulls, and thus, death. In fact one of his brews is called "The Reaper." This beer is actually quite pleasing to the palate. He has not yet begun bottling his beer thus there is no label attached just yet. Still, I wonder what kind of affect this name might have on the beer itself. Even intention can be infused into the things we consume.

I have also seen bottles of wine with demon-looking creatures on the labels and with names containing the word "devil". Many of these wines have highly artistic and interesting looking labels. But despite the appeal of the way they look I often wonder what those words, with negative connotations, or flat-out demonic names, might be doing to the structure of the water from which the wine is made and what the wine might do to a person who drinks it. Considering Dr. Emoto's work it might be best to avoid such wine.

The monks at a nearby monastery make some of my favorite craft beers, Monk's Ale, Monk's Wit, Monk's Dubbel Ale, and Monk's Tripel. The beer has a label, containing the words, "made with care and prayer." This statement alone has the ability to create a positive intention in the beverage. Dr. Albrecht encourages the infusion of positive messages and intention into anything we consume, including beverages. Blessing our food and drink prior to consuming has great potential for encouraging health.

The same is true with regards to the labels on other types of alcohol commonly referred to as "spirits". This includes such drinks as vodka, tequila, whiskey, and other similar hard liquors. As it is they are called "spirits" for a reason thus infusing them with negative words could enhance this alleged "spiritual" effect and not in a good way.

These things might sound silly but understanding how energy manifests within words and how powerful energy truly is one would do well to ponder on this and maybe even pay attention to subtle cues when drinking such beverages. Intention is also a huge issue. What was the intention of the manufacturer? What is your intention just prior to drinking? How does it make you feel? Agitated? Happy? Sinister? It is all about energy.

We all know the body treats alcohol as a poison thus its very consumption is detrimental to health. Imagine consuming alcohol that has been infused with negative energies and how much more detrimental it might be. Conversely imagine having the power and capability of counteracting not only negative energies but also even the toxicity of the substance simply by infusing it with positive messages and blessings. In fact Dr. Emoto discusses this very thing in one of his books. A group of individuals were able to decontaminate a large body of water simply with positive, loving thoughts. It might seem unbelievable but, again, energy is pervasive and profound. Our intention and thoughts have incredible power to overcome quite a bit.

Such concepts as infusing positivity with harmonious frequencies, whether by the intention of words or the frequency of sounds are not new. Studies have shown that fetuses exposed to classical music have a higher IQ than those exposed to no music at all. The evidence is there. Now it is time for all of us to put it into practice. For more on this topic conduct a search for any of the books by Dr. Emoto. You might be surprised and pleasantly enlightened.

Chapter 19

Invisible Frequencies
Just Because You Can't See It
Doesn't Mean It Isn't There

Before I move on into the next chapters it is appropriate that I delve a little deeper into the concept of frequencies and their invisible nature. In order to really understand energy it is necessary to put it into a certain context so you can truly visualize the invisible nature of energy. I know that sounds outrageous and an oxymoron but bear with me here. When you have finished reading this chapter I believe you will grasp what I am trying to say.

The term "frequency" relates to the frequency of vibration per second, as it exists within a given thing, which is measured in Hertz. As I previously stated, energy exists at various ranges of frequencies, including ELF (extremely low frequencies) to UHF (ultra high frequencies) with RF (radio frequencies), and others in between. And within each of these groups exists numerous individual frequencies. So, conceivably there are billions of frequencies beginning with nano or micro-Hertz, up to giga-Hertz.

On earth each "thing", alive or not, exists within a certain range of frequencies. You see, each *and every* "thing" is made up of atoms and molecules; there are no exceptions. The chair you sit on, the pen you use for writing, your computer, your coffee cup; everything is composed of various atoms and molecules held in a certain structure. Even the color of each of these things is only visible due to energy. It is the energy that acts like the glue to hold these things into their various shapes.

The same is true for living organisms. The energy within living beings is just more active and complex. So far I have not delved too much into the actual measure of the energy within the body itself in terms of frequency, however, it has been known probably since the beginning of time and well studied scientifically that our bodies function optimally at certain frequencies and when there is some sort of distress within the body, including illness, the frequency changes, as in it goes "off balance".

For example, if the sound wave produced by the frequency is not harmonious and is imbalanced it is unpleasant to the auditory senses. Speaking of sound waves, music itself has been measured according to its frequency and the most pleasing music exists at certain specific frequencies. When those frequencies are thrown off balance or in disharmony it is audibly noticeable.

That frequencies have the potential for healing or destruction is well known; take, for example, opera singers who are able to sing at such high pitches they are able to shatter glass. This concept will be discussed further in the next chapter but for purposes of this discussion, I would like to illustrate how when such an event occurs you are unable to visibly see what causes the glass to shatter. The same is true with regards to animals that can hear different frequencies than humans. There is a reason for all of this, some of which is because animals are very intuitive, which is critical for their survival.

The discussion of energy with regards to animals and the frequencies to which they are sensitive could go on for days; the point is that life operates at varying frequencies and all objects function optimally at certain frequencies, and living organisms are sensitive to certain frequencies.

The signal coming through your television also comes through based on a certain frequency and the same is true for audio. When it comes to radio, the sound exists within the range of radio frequencies (RF). You must tune the dial to a particular "station" (frequency) in order to hear the sound passing through that frequency. If you have not tuned the dial to the particular station, does it mean the frequency is not there? Absolutely not, it just means you cannot hear the sound coming through on that particular frequency. The same holds true for television. Back in the day the televisions picked up signal (images and sound) within the UHF and VHF frequencies but the particular picture you wanted to see came through on a single frequency within that range. They were called analog signals. These days the image and sound comes through digital signals, which utilizes satellite technology, which uses outer space.

This type of phenomenon is well known and greatly studied by quantum physicists. A simple search on the Internet yields thousands of results relating the so-called "spiritual" realm to science. This is where you have to really try to conceptualize the whole energy, spirituality, multi-dimensional planes, and science idea. Most people who believe in Intelligent Design often ponder the meaning of the spiritual realm but rarely do they consider its relation to science.

Before you can truly conceptualize this you must have some understanding of the energetic aspect of all beings otherwise this incredibly interesting information might be lost on you. This is why I provided an in depth discussion on energy and the way it operates within the body.

So now that you have an idea that we are actually less dense than we look and it is mostly a perception based on energetic frequencies of our existence you could begin to imagine the possibility of other planes of existence – the so-called spiritual planes. Since there are millions of frequencies, some of which are only audible to certain creatures then imagine that some of those are also only visible to certain creatures as well.

Just because most humans cannot "see" things functioning at certain frequencies does not mean they are not there. Consider animals when they sense danger or bad "energy". It is entirely possible that they can actually see something that physically exists only on another energetic plane or within another dimension or range of frequencies. This is starting to sound very strange but hear me out.

As an example I'll tell you a story. When my dog was just a puppy, around six months old, it seemed she sensed an invisible energy in my room. Of course it was invisible to me but might have been visible to her. I remember her staring and growling toward a desk I had in my room, with a Bose radio on top of it. Shortly after this happened the radio turned on "by itself." I still don't believe this was simply an electronic glitch. It had never happened before and it never happened again after that, the machine did not malfunction. I am convinced she saw something there.

Think about it this way. If I can hear music in my kitchen from a radio, the sound of which is actually being played miles away, but it has to be tuned in to a certain frequency in order for my auditory nerves to pick up on the sound, then why would the same not be true in terms of visibility of a physical being existing at an entirely different frequency or range of frequencies?

Could it be that we could create a tool, such as a radio or television, which would allow us to "see" a different type of energy passing through our space at a different frequency than our eyes or body alone are capable of witnessing? Those involved in Quantum Mechanics and Physics are actually studying these things presently. The technology might actually exist, we might just not be aware of it.

This all being said I would like to present to you an idea that scientists are already discussing. The idea is that "ghosts" or spirits actually can be seen in physical form, however on a different plane or frequency than our human senses are able to visualize. Like the television or radio frequencies we cannot see or hear but become visible and audible once our machine has tuned into the correct frequency it is very possible, and in some cases proven, that there are living, spiritual entities all around us and only those who are able to use their own bodies as machines or tools to tune into the frequencies of those entities are aware of their existence.

I believe most people have had supernatural or spiritual experiences where they were able to see, hear, or even just feel the presence of another entity. It is entirely possible that the particular sense allowing that experience happened to be tuned in to the correct frequency at the right time. Thus, these "spiritual" experiences that some people tend to have more than others are actually footed in science.

Could it be we are all capable of sensing or even seeing entities that exist on different planes, at different frequencies, we just are unaware of how to tune in to that particular sense? What if every single dimension and plane exists within the same "space" just at different frequencies and ranges of frequencies? Just as we are able to practice understanding our intuition, could this be another sense, perhaps a seventh, eighth, ninth sense, or beyond, we could also practice in order to tune into other frequencies? Consider the possibilities.

Chapter 20

The Royal Rife Invention

After receiving my undergraduate degree in chemical engineering and before I made the decision to work toward my master's degree almost twenty years ago I was about to start working on a project where I would use an ultrasonic bath to deeply clean the debris from items containing radioactive materials. The contents of bath itself were only of minimal concern; it was the frequencies used with the bath and materials that was the focus of the experiment. The concept of the experiment was to use vibrational frequencies along with the bath to remove the debris. I was to determine the lowest frequencies of vibration necessary to get a good cleaning of the items. Alas, as it happens frequently at national laboratories, funding went to another project and I was never able to work on this one.

Little did I know at the time that the very same concept had been researched, and an incredible invention made, decades before by a man named Dr. Royal Raymond Rife, although his invention would act to accomplish something similar within the body and on living pathogens. The area of research is known as electro-medicine. It is rooted in energy healing except instead of working on the vibrational frequencies of the human body to facilitate healing, as most energetic healing modalities, it works on the energy of the pathogens toward destruction.

Typically when we think of energy healing we think of the human body and how to change things with regards to our own frequencies. We now know the earth has its own frequency at which it oscillates and that its frequency sometimes varies due to geomagnetic activity. We also know our bodies require connection to the earth to absorb these very frequencies. In order for the earth to be energized as such it requires lightning strikes and purportedly there are roughly 1,000 lightning strikes to the earth per minute around the world.

Most energetic healing modalities incorporate this information and work toward healing using our own energy; however, with regards to the Rife treatment it is actually the pathogens' frequencies that are targeted. Dr. Rife was able to develop a laser, which would operate at the specific frequencies upon which the pathogens functioned or lived. In this way he was able to actually disrupt their frequencies to the point that they could be destroyed and expelled through the body's normal excretion pathways.

Much like the ultrasonic bath I was tasked to research the Rife technology operates similarly. Instead of cleaning debris from materials within a bath, the technology "cleans" pathogens within the body. Unbelievable? Maybe, but similar technologies are being used at this moment to help people rid themselves of cancer. Of course this is not being used in the mainstream, as anyone who bucks the system with ideas and understanding that the body might possibly cure itself without the need for toxic pharmaceuticals to do so is deemed a quack.

By typical mainstream thug fashion, Dr. Rife was ridiculed and tormented for his invention. The whole story can be found by a simple search online. Like many other brilliant medical doctors who figured out ways to help the body heal itself without the toxic effects of the poisons pushed by the pharmaceutical companies this man was driven out of the country and fully traumatized for daring to think outside the box.

In studying natural healing modalities, energetic or otherwise, I have come to realize it has all been very concerted and when a doctor is deemed a "quack" by the mainstream they likely have something positive to offer. One web site, which disguises its bias as much as it can, is called quackwatch.org. It is an Internet shill web site meant to deter people from looking at legitimate treatment options by vilifying those who would dare to go against Big Pharma.

The way it disguises its bias is by printing legitimate articles about people - naturopathic or medical doctors or laypeople - who tout treatments that have not been studied or that really provide little to no help. These people, referred to as "Snake Oil Salesmen", do exist and they are the ones who give the legitimate treatments a bad name. My cynical mind says that these people exist solely to discredit the majority. After doing as much research into these things as I have you would not blame me for believing this way.

Anyway, the point again is that just because you cannot see energy doesn't mean it isn't there and it especially doesn't mean something is not happening within your body, good or bad, based on your own energetic frequencies or those of the microorganisms that have made an unsavory home in your body.

So, the takeaway from this chapter is that, just as you are an energetic being, so are all the billions of microorganisms that have the ability to either help – think probiotics – or harm – think every other pathogen – "bad" bacteria, viruses, etc.

Chapter 21

Emotional Freedom Technique Clearing Negative Energies Through Tapping

There are many healing modalities and techniques I have previously learned of but only recently began studying intently. One of these is EFT or "Emotional Freedom Technique". As with many other things in life sometimes we are not ready for certain information the first time it is presented to us. In essence, information presents itself when most needed or, in another sense, we are drawn to certain types of information when it is most needed. For me this particular topic happens to be such type of information.

When I first heard about EFT I could not appreciate - or understand - the power in this technique. Although I had some idea back then of harnessing energy in a positive way for a number of things it was not until the past few years that I have taken a great deal of time to study and try to truly understand this invisible phenomenon we call energy and why it is so important.

Now that the concept of energy should have been somewhat demystified you will surely appreciate the simplicity of this technique and might consider looking further into it so you could also benefit from clearing energy blockages, which might be causing various issues within your body and mind. These issues can include physical pain, depression, and even financial failure. Thus, EFT or another similar modality, including doing affirmations, has the ability to reverse all of those things. It has the power to relieve physical pain, release depression for true happiness and even help you find true, lasting success in your life.

This technique is an offshoot of another, called Thought Field Therapy, or TFT. A psychologist by the name of Dr. Roger Callahan developed TFT and, sometime thereafter a man by the name of Gary Craig modified the technique to make it simpler and called it EFT. The efficacy of both has been touted, particularly when the procedures are followed closely. However, purportedly there is more room for "mistakes" or diversion with EFT, as great benefits are seen even when not using a specific routine or protocol with this method.

Earlier in this book I discussed my chiropractor, Dr. Storkan, who uses kinesiology in both the diagnosis and treatment of his patients. Namely, he has used it numerous times on me when treating my allergies as well as other issues. The way he uses the kinesiology, or muscle testing, is by asking my body specific questions regarding the issues. At the start of each session, Dr. Storkan has his patients place one hand on top of the head, palm down, while placing resistance on the opposite arm. He then does the same thing with the patient placing one hand on top of the head, palm up.

If the patient is unable to resist his force while the palm is face down it means the patient's "wires", e.g. nerve impulses or messages, are crossed. If the patient is able to resist his force while the palm is face down, and unable to resist the force while the palm is face up the patient's wires are fine and he can proceed with the treatment. What does this mean? Well, first, when it comes to "wiring" of the brain, it means just that: the neuronal connections and synapses within the nervous system through which the brain communicates with the nerves and then the cells of the body.

Dr. Storkan uses this very simple method to ensure he can trust what the patient's subconscious is telling him during diagnosis and treatment. Essentially, it is a way to ensure that the patient's yes means yes and no means no. If not then the treatment would be for naught.

In many cases when the patient is under stress of some sort the wires are almost certainly crossed. This has happened to me on more than one occasion. Dr. Storkan is able to pinpoint my issue as being related to stress and also the organ or organs affected by the stress.

Several years ago I experienced a traumatic incident that triggered heart palpitations. The palpitations came frequently and although I was and have been fairly healthy they startled me. After several weeks of dealing with this condition I went to the mainstream medical doctor to have my heart checked. I rarely see a medical doctor for any issue therefore it was out of my element but I felt the need to find out what the mainstream medical system might come up with as far as a diagnosis, using their typical diagnostic tools. In my view, using mainstream diagnosis can provide insight into immediate danger thus I felt it was a way to determine, quickly, if I should be worried.

Although I am generally apprehensive about mainstream treatment, particularly in the form of pharmaceuticals, I do believe mainstream medicine serves an important function particularly with regards to emergency situations and I was concerned this might be such a situation. After having gone through the standard tests such as an echocardiogram, EKG, and carrying a Holter monitor for twenty-four hours and determining I was not in imminent danger I made an appointment with Dr. Storkan.

Upon first going through the initial diagnosis he correctly predicted or, more appropriately stated, diagnosed, I had been under a great deal of stress. He pinpointed the stress as directly affecting my adrenals, which is typical. After being treated for the anxiety, strengthening the adrenals, and using trigger point therapy he informed me that I should feel better within the next few days. Indeed the heart palpitations ceased and I have only had one sporadically since, with the latest incidence happening more than a year ago. Of course, because of the stress my own wires were crossed and it was not until this was resolved that he was able to properly treat the anxiety. This can also happen even if the issue is not directly related to stress but the patient happens to be dealing with some sort of stress.

The method Dr. Storkan uses to "uncross" the wires is a very simple EFT tapping sequence. It is the very first point taught using this technique. The patient must tap the outer side of the hand using the index, middle, and ring fingers of the opposite hand. While tapping the patient states, "even though I feel miserable, I deeply and completely love and accept myself," three times.

As previously discussed, once this is complete Dr. Storkan then asks the patient to place her hand on top of her head, palm down, and, because of the EFT tapping the polarity should be proper such that the opposite arm is able to resist his force. Every time this sequence or tapping exercise is completed there is always complete resolution of any reversal of polarity. Once this is done his treatment can continue as usual. In another sense this is a way for the person to connect with herself with love. Self-love is one of the most important tools for healing and indeed a successful life.

As with this particular exercise many people have found great success using EFT for other issues including pain relief and even finding success in their lives. With regards to energy the idea of EFT is that, by tapping on certain meridians, the energy flow in the body is optimized such that any blockages or redirects of that flow are corrected.

The statements used can be anything from this technique for "uncrossing" your wires to whatever you wish to achieve. By stating, "even though I feel miserable I deeply and completely love and accept myself" you are embedding the message into your subconscious mind not only giving it permission to direct you toward success but also realigning your conscious and subconscious mind toward the positive.

We tend to limit ourselves with our own negative thoughts, and, unfortunately, they are usually most directed at ourselves. We are our own worst critics and thus our own worst enemies. We know, deep down, we are responsible for our own situation or destiny in life. When things do not go our way we, on a subconscious level, blame ourselves and many times we fail to forgive ourselves. In the meantime we generally have no idea this is why we feel so miserable. It is my belief that at the root of our misery, whether it is physical or emotional, there is a lack of forgiveness of self and general lack of self-love, which is essentially one and the same.

The statement reminds us we are worthy of love and that we are loved and, first and foremost, it is a love of self. It begins with acceptance and forgiveness of self. Sometimes this is difficult to achieve because by consciously admitting there is something for which we must forgive ourselves is also a conscious admission we have somehow failed. Acknowledging failure is one of the most difficult things for humans to do, thus by tapping you are able to activate certain meridians, clearing the blockages along those lines without requiring your conscious mind to accept the prospect that you might not be perfect. When you tap while making these statements, whether your conscious mind believes them or not, you are still sending the message to your subconscious mind, even more so than if you were to simply just make the statements alone.

Other statements toward goals and success can be used as affirmations themselves. For example, if you wish to have the job of your dreams making a comfortable living you could tap while constantly stating the goal as if it has already been achieved. "I am living my passion of helping others through improving their lives and I make XX dollars each year," for example. Stating your intentions, goals, aspirations, and dreams both on paper and out loud is powerful on its own, but stating them while tapping meridians is actually embedding these messages in the electrical wiring of your body, the nerves, and sent directly to your nervous system (brain) and subconscious mind.

I have used this technique very successfully to quickly change my mood. When I have been frustrated with a situation I simply conduct a tapping sequence through which I repeat certain statements such as, "I feel at ease, I have patience and am at peace." I have also used it when I procrastinate or feel a lack of motivation and it successfully gives me an energetic push to work toward my goals.

This technique has been successfully used for many issues, not the least of which is to foster more effective healing. The possibilities of the uses for tapping are endless. Studying the resources on this subject, included in the Appendix of this book, would prove highly beneficial, as use of the technique is capable of addressing many issues.

Chapter 22

Network Spinal Analysis
Reorganizational Healing

The basis of my philosophy, and indeed the anchor for all of my work, is that the body is self-healing. There is no such thing as a "cure" for anything because the body is capable of curing itself. As such there is also no such thing as a "healer". There are only tools to facilitate healing within the body, which includes the practitioner himself using his own gifts and skills as the tools or himself as the conduit. Of course there are nutrients and other practices which can help speed along the process and, conversely, toxins and pathogens which can stunt that process or make the process impossible altogether.

The question then becomes, "is *my* body capable of healing itself and, if so, is it healing itself right now?" In the physical realm there are different ways these questions could be answered. For example, it is well known that the body heals when it is at rest and, in particular, when you are in a deep sleep. This is when melatonin is created. As discussed before melatonin is both an antioxidant, thus helping the body to heal by neutralizing free radicals formed from excess inflammation in the body, and also a sleep aid thus helping you get into and stay in a deep sleep for the necessary amount of time.

But does energy play a role in ensuring optimal healing? Is there a way to achieve optimal healing through the use of energy? A chiropractor by the name of Donald Epstein created a healing modality through which energy is the main component. His method is considered reorganizational healing rather than restorative healing. While many other chiropractors and other natural practitioners are able to facilitate healing using various energetic techniques, some of which I have already discussed, they mostly typically involve restorative healing, meaning they are "correcting" a problem.

With Dr. Epstein's modality, which uses a concept called Network Spinal Analysis (NSA) to determine baseline nervous system function and also to track progress, the idea is to "retrain" the nervous system or entrain the spine with the proper energetic frequencies as intended.

The nervous system begins with the brain, then the nerves and spine, which communicate messages to all cells in the body. This "retraining" is intended to essentially re-educate your body on how to deal with adversity, including emotional conflict as well as injury, such as traumas resulting from accidents or illness.

A chiropractic session is called an entrainment and the practitioner uses very slight contacts with your body to open gateways into the spine. Once he is able to access the energy of the spine he is able to redirect the energy such that your body begins to heal more efficiently.

From birth we begin developing habits of dealing with life. Sometimes we consciously or very slightly subconsciously create these habits and other times they manifest on their own. This has to do with the way our bodies utilize the energy surrounding an event or what is mobilized within the body during the event. Injury and other stressful situations cause a reaction within the body. Sometimes it manifests as tension wherein there is a pocket of energy just sitting there, wreaking havoc. What we do with this energy is what determines whether we create a healthy or unhealthy pattern of responding.

For many years I felt a dull ache just above my left shoulder blade, particularly when I was stressed out. For the most part I ignored this pain, until shortly after my father passed away and I decided to get a massage. I knew a knot had formed; however, I did not realize how long it had been there or how bad it had become until the massage therapist told me it felt as though the knot had become fused to the muscle. As a result I began getting massage regularly for many years; however, unfortunately, the knot never went away with massage.

As a side note I would like to point out how the knot became greatly uncomfortable upon the passing of my father, whose tension in the days preceding his death was affecting my own physical body, as I was living in the same house at the time. Beginning exactly one week before his death right around seven o'clock in the evening each day I would suffer a horrific headache. It was not a migraine headache but the most excruciating tension headache I have ever experienced. It was not relieved with medication and all I could do was go straight to bed. I believe my body had been feeling the tension of my father's impending death, as the headaches ceased on the day he died and I have not suffered one since. I feel this illustration supports the discussion within this book regarding our connection with others and the collective energy of all of mankind.

Another interesting point I should bring up with regards to this knot, which wanted to make a permanent home on my body is, even after it went away, and up to this day its location has served an important function. I am able to sense an infection within my body because of a certain sensation of weakness and dull pain in that location. Without fail, when I feel pain in the area it is always due to the onset of infection. While discomfort and pain are usually unwelcomed I find it interesting, and even a gift, that understanding my body in this way has allowed me to address the infection as soon as it comes on so as to clear it out within a few short days. This has included strep and other infections. Such acknowledgment and acceptance of pain and discomfort speaks to appreciating our bodies for all they do, including sending messages of pain to facilitate healing.

As for dissolving the knot, enter traditional chiropractic. I always knew what it was but never felt the need to see a chiropractor, as I did not see my issues stemming from any misalignment of my spine. Little did I realize that all issues essentially come from a misalignment of the spine. I began receiving regular adjustments, at first twice a week and then down to one, for a few weeks. It took some time but eventually the knot went away, thanks only to the chiropractic treatments.

After this I regularly saw various chiropractors for other issues, especially when I would throw out my neck or back, or when it just felt like my back needed to "pop." In addition to treating me for allergies Dr. Storkan, who is also a chiropractor, has treated me for many other issues, including musculoskeletal problems. Such issues can be treated energetically with success even in a restorative sense.

For example, I had suffered from low back pain for nearly two years. This pain had seriously hindered my exercise and running routine, which I had all but abandoned due to the pain. I had begun receiving regular adjustments from my traditional chiropractor in the hopes of taking care of the issue. His technique is not related to energy but typical spinal adjustments using the Gonstead technique. This provided some relief but very little, as the pain was simply too much for traditional chiropractic.

It had gotten to the point where I would feel a sensation of numbness in my legs at times, which concerned me greatly. Such symptoms generally prompt mainstream doctors to recommend surgery, which can be extremely damaging to the entire nervous system. Many people do not realize how such treatments as surgery are also only sometimes necessary in emergency situations. Even with regards to musculoskeletal issues and nerve damage the body is still capable of healing itself with the proper assistance.

Opting not to entertain the thought of many people ultimately going through surgery for this type of pain I finally made a decision to see Dr. Storkan for this issue. He went through his typical diagnosis using kinesiology then a series of treatments using his chiropractic tool, which works to clear energy blockages; he also performed trigger point therapy, which is highly effective. After he completed the treatment he said, "You should feel fine within three or four days. If you don't, give me a call."

I thought, "Ha! Sure pal, I've had this pain for two years, I'm pretty sure I'll need more than one treatment from you." I then asked what he did and what was wrong. He stated my lower body muscles were all "blown." This was strange and I had no idea what he was talking about. I asked, "So should I do strengthening exercises or something?"

He responded, "No, the muscles just weren't working the way they're supposed to. I cleared up the blockages so they'll function again." I found this fascinating if not just a tad bit incredulous and unbelievable.

Essentially, the muscles were not receiving the signals from the brain so as to function as they were intended. The muscles simply weren't working, thus forcing my lower back to take on the entire load. Lo and behold, the pain was gone within a few short days and I did not have to return for additional treatment. This was several years ago and only occasionally do I have lower back pain, which typically resolves itself quickly. Although the problem was easily and quickly resolved using energetic healing, it was still a restorative rather than reorganizational modality.

At the time I assumed this is how it would always be and I would always need a chiropractor to adjust my spine, using conventional or even energetic techniques. Then I found out about NSA, which is offered locally at a place called The Scher Center for Wellbeing.

Because the nervous system is the foundation of everything it is necessary to have an optimally functioning nervous system in order to ensure the body is doing what it needs for health and healing. The body is our greatest asset, and it can make or break us depending on its health. The root of all health is the brain, which connects to the spine, which connects to the nerves, which connect to everything else within the body. The nerves can be viewed as the communication network of the body; the neurons that pass through the nerves can be seen as the messengers sending important details from the brain into the many cells of the body. It is these neurons that tell your lungs when to inhale or exhale; they tell the heart how much blood to pump; the kidneys how to filter toxins; and your feet how to walk.

Imagine that any misfire of these neurons, or any break in that communication, could cause some serious issues within your body. Over time we can cause changes within that communication network in various ways for example, by favoring a part of the body that might be in pain. If our bodies have become accustomed to doing such things we can begin a distortion process within our spine, which thus distorts the messages from the brain to the cells.

Many of these changes are subtle. For the most part we do not even notice we are doing this until, one day, we look in the mirror and see a hunched back (kyphosis) or a twisted spine (scoliosis). Without realizing it we are causing ourselves more pain and suffering by twisting the body into an unnatural shape to avoid pain or discomfort. Sometimes we are able to manage the pain with toxic medications, but consider that because we are masking the pain with such medications, we are not paying attention to what our body is truly going through and thus creating more damage. On top of this, as a result of the distortion of our spine, the reality of our world begins to distort as well.

Consider those who live with chronic pain, whether or not it is being "managed." These people are generally very unhappy, many create drama for themselves and others; and, as the adage states, "misery loves company." The spinal distortion of a single person can negatively affect a whole family or even community.

Being entrained involves using these gentle and gradual contacts, which tap into the energy along your spine through certain gateways with the goal of ultimately accessing trapped energy. The spine soon begins to remember what it once was before we allowed distorted messages and began creating unhealthy habits of storing stress in various areas of the body. Over time you can visibly see the posture improve. Again, these changes are subtle. For it took many years to become so distorted, one cannot expect this to be undone overnight.

When I first discovered NSA I had already begun seeing my head lunging a bit forward. It was nothing major but I noticed it when I would look at my profile. I felt a sort of hump on the upper part of my back, below the base of my neck. For me it was becoming a bit unsightly. I had no idea how I would ever deal with this issue. I assumed it would just always be there and hoped it would not get worse.

Right around the time I discovered The Scher Center for Wellbeing in Santa Fe, I had been dealing with some emotional issues resulting from death, illness, and drama in my family. These issues had compounded to the point where I felt I could not deal with it all anymore. It wasn't a feeling of wanting to give up on life or anything like that, but a feeling of such utter overwhelm I was ready to release responsibility for various things, including caring for my mother, who had been ill since my sister passed away. I had never experienced this feeling of overwhelm before at any time in my life. It was as though the pot that had begun to boil months before was about to overflow, or perhaps was already overflowing.

Knowing what I knew about energy and because of how I had been feeling, I knew whatever healing I needed for my emotional state had to be energetic in nature; I did not know what it was, however. I had already been trained and certified in Innerwise® but I was in no condition to tap into my own intuition to help myself, as my emotions and stress were already clouding my own judgment and my subconscious mind was receiving the confusing messages of which I previously spoke. I had become my own worst enemy.

The way life works sometimes is quite interesting, when you know and feel you need something it tends to show up in the most unlikely way but right when it is most necessary, as I previously mentioned. For me it was a Facebook invitation a friend of mine shared to an event called, "Wine, Women, and Chocolate," three of my favorite things. It was also free. The event consisted of a few women presenting on various topics; it was held at The Scher Center with Dr. Judy Scher as one of the presenters. She explained the concept of NSA, reorganizational healing, and gave a demonstration on one of the practice members. It just clicked and I remember thinking, "this is exactly what I've been needing."

Needless to say I have been under care for over a year now and so many things have happened and many changes have occurred. It has been a bit of a roller coaster but I feel much more at ease and at peace. Actually, the changes have been profound warranting an entire book on the subject. Because retraining the body toward optimal self-healing is a concept encompassing a great many things it is difficult to articulate the power in correcting habits that once led to the distortion of the spine. Figure that reversing such distortion not only has the ability to redirect energy flow in the body to allow cellular improvements but it also has the power to change our entire perspective on life.

One of the things the doctors iterate and reiterate is that we each have choices. We can become victims of our adversity or we could use it as an opportunity. In an emotional sense, there is not much more I can say about it, as it has to be experienced to truly understand its power. On a physical level it is much like the treatment I received from Dr. Storkan, which completely resolved my lower back pain within a couple of days. Entraining the spine toward optimal healing has the ability to clear many blockages, which might not be quantified through physical pain. Retraining the body to relay messages in an optimal fashion has the ability to allow the body to heal even the dastardliest of illnesses, those that might not even be diagnosed or recognized.

I will say this: Each potential patient is scanned initially for temperature, electricity, stress level at rest, and posture, the latter of which is a visual image. Patients are then rescanned after being under care every six weeks. The hump I had developed or was developing at the base of my neck has greatly improved. My chin has inched its way back, and my head weighs less – as a result of its positioning upon my spine – than it did when I began. Not only am I much more aware of my posture and breathing, I feel much more open to opportunity and much more capable of dealing with challenges. The idea is for the spine to adjust so that, eventually, it will have a capability of adjusting itself during injury or stress, as it was intended.

When I had my first follow-up scans I remember having been dealing with a lot of unnecessary drama. Of course drama is always unnecessary but at this time I was dealing with something fairly absurd. A random individual I had not seen in many years, and who was never really anything to me anyway, suddenly decided to harass me. It was quite surreal. I remember feeling like I was under attack even though the issue should not have been a big deal. So, after my first scans were complete and we went over them, the person at the center showed me the images of my profile. I had mentioned how I was feeling then she pointed out the way my head appeared to lunge forward and said, "No wonder you feel like you're under attack; your body has the fight stance."

This was pretty profound for me. It drove the point home that however our spine is distorted our reality will be distorted in the same way. Logically I knew this. I realize that anytime someone creates problems, which enter my space, they are not my issues and I can choose not to let them affect me, however I had somehow forgotten this. It took seeing it with my own eyes to realize it was entirely possible the way I had been feeling was simply due to my body's unnatural reaction to stress, which had compounded over the years and especially in the previous year.

Needless to say that since I have been under care my outlook has improved greatly, my productivity has shot through the roof, and even my creativity and path forward in what I hope to achieve have solidified whereas before I felt as though I was aimlessly wandering, just spinning my wheels. I am able to react to challenges and conflict in a much more effective way and physically I feel better than I had in years but mostly I feel much more capable of dealing with life's challenges. It would not be a stretch to say the completion of this book is partly due to the incredible transformation I have undergone at The Scher Center. As I stated early on in this book, there are no coincidences.

As part of the practice, members are directed and taught to connect with oneself through breathing, a practice called Somato Respiratory Integration, or SRI. This practice has proven invaluable to me and has resulted in many breakthroughs. Pain and discomfort, physical or emotional, is welcomed as it is through our body that we are able to listen to the subconscious mind, as it communicates its needs. The practice of breathing through issues and feeling the body's response provides deep connection, acknowledgement, understanding, and ultimately acceptance. This also is a practice of self-love.

Throughout this book, if you pay attention you will see a common theme. This theme always goes back to forgiveness and love of self, which has the power to change the physiology of a body and the emotional state of the mind. In Chapter 24 you will learn how these concepts can affect the world as a whole, helping you to feel empowered by understanding your role in the grand scheme of things.

Chapter 23

Self-Affirmations
A Physiological Connection
Between Thoughts and Goals

The message of this book I most wish to convey is that we have the power, within ourselves, to receive whatever we seek including health, happiness, love, and even wealth and abundance in every way. For this reason it is important to discuss the scientific connection between thoughts and their role in mental health and behaviors. Although I previously discussed affirmations briefly I believe it is important to have an understanding of their true power.

As self-affirmations have gained popularity and many people are using them to effect positive change in their lives there are more studies being done to determine exactly how they operate within the mind and body. Science might not ever be able to completely unravel the mystery but it is doing a pretty good job of shedding some light.

Of course there is always a spiritual connection because faith has a profound effect on wellbeing. Many people who suffer from serious illness are able to get through the experience much more easily when they have faith than when they become victims of their circumstance. I have always felt doctors do their patients major disservice by giving them a deadline of death. When a doctor says, "you have six months to live" it is incredibly damaging to the psyche. In fact these words seep into the subconscious mind to where the subconscious seems to ultimately make it happen. Counteracting this type of message is critical and some people have mastered the technique.

It is like anything else. When a person constantly states, "I'm in pain," or, "I feel miserable," etc. those messages eventually become hardwired to the subconscious. The subconscious either becomes confused, causing the very turmoil to which the person dooms himself, as these messages are not conducive to success. Or it makes the negative goal come to pass simply because the subconscious mind is responsible for directing the body in line with a particular belief system. Essentially the subconscious mind thinks, "I better make sure there is pain in this body, because that is how it is supposed to be." Conversely when we constantly say, out loud and especially in writing, "I feel good," or, "I'm healthy," or other positive ideas, the message becomes hardwired into the subconscious such that the subconscious mind then says, "I better make sure this body feels good"; moreover, this message is one of success and thus creates a situation of harmony and balance.

It really is not that difficult to grasp. This is why I previously stated it is incredibly important for us to protect our space and the messages we allow into our subconscious minds as well as the energy of others. This includes the messages and energy we receive from doctors. Some people put a great deal of importance on what their doctors say; it is almost like a religion in and of itself. They see their doctors as gods, with infallible predictions and recommendations.

I am aware of people who were told they had cancer for many years but once they receive the diagnosis, and the statement that they have "x" number of months to live, suddenly they are on their deathbeds and, guess what, they only last those "x" number of months. Words are powerful.

What if you could create a different reality? What if you had more control over your health simply by what you allow into your mind and what you speak? What if you could turn things around simply by reprogramming the negative messaging with which we have been bombarded? Well, based on what I have researched you absolutely could.

The autonomic nervous system has two divisions – the sympathetic and the parasympathetic system. The sympathetic is responsible for your fight or flight reaction while the parasympathetic is responsible for healing within the body. It is important that the function of these two divisions is balanced so that you will not only heal while at rest and, especially, during deep sleep but also so that you will have the ability to properly react to a life-threatening situation. It is unhealthy and even dangerous for one or the other to dominate. Again, our bodies seek balance.

The sympathetic system operates from our primal brain at the base of the skull, the hippocampus. It is the primitive part of the brain responsible for base survival. It is also the location where memories are stored. The prefrontal cortex (PFC), just underneath the forehead is considered the "higher" brain, which is responsible for rational thought, behavior and from where the more complicated messages to the body for healing are sent. Obviously the brain requires a great deal of oxygen just to function and for the prefrontal cortex to function optimally it is critical that it receives as much oxygen as necessary. We know deep breathing helps to trigger the parasympathetic system.

Unfortunately when the spine is distorted in any way some of that oxygen will likely not make its way to this part of the brain. This is partially why deep breathing techniques are so beneficial for healing. Through the reorganizational healing modality, as discussed in the previous chapter, practitioners spend much time focusing on "Somato Respiratory Integration," SRI, to aid in the healing process. This is a practice through which the patients place their hands on three separate locations of the body while breathing deeply and focusing on any discomfort within the body, physical or emotional.

Deep breathing sends oxygen up the spine into the brain all the way to the PFC provided there are no blockages. This then activates the PFC, which is responsible for rational thought and a higher level of intelligence. Activating this area stimulates the parasympathetic system to encourage healing. It also encourages rational thinking in a fight or flight situation such that the person is able to respond properly.

What does this have to do with affirmations and subliminal messages? It is known that subliminal messages can make their way into the subconscious mind potentially resulting in unwise or unhealthy decisions. These messages can either confuse the subconscious mind to be in disharmony thus causing turmoil within the body or, they have the ability to convince the subconscious mind to direct a person toward making certain unhealthy choices and engaging in certain behaviors. The advertising industry is very lucrative because it has determined how to embed messages into the subconscious mind, which can effectively direct a person to make certain purchases and engage in certain behaviors, regardless of whether they are healthy or not.

Despite its efficacy and the fact that companies pay millions of dollars to embed subliminal messaging into their ads it is difficult to locate scientific studies on how exactly they work. Could it be that there are genuinely few studies on the subject or could it be that these studies are somewhat hidden from the general public? It would unbelievable that scientists have not mapped the brain during the viewing of such ads so as to justify the industry's investment therefore I am inclined to believe the latter.

In my Communications 101 class in college we had a lecture dedicated to subliminal messaging, which was very eye opening. The fact that it was part of the curriculum implicates its prevalence in society. We would do well to understand or at least be aware of its use either to protect our subconscious minds or to use the technology for self-improvement. My professors gave a full presentation with images from magazine, newspaper, and television ads containing various subliminal messages.

Subliminal messaging can be embedded visually through photos or verbally with words, and particularly through repetition. There are many books and even the book and movie "*The Secret*," which is a compilation of interviews, that address this concept. As previously discussed Napoleon Hill wrote *Think and Grow Rich* to convey information on this topic. I believe the reason affirmations work so well is because the practice consists of the conscious mind sending messages to the subconscious, essentially rewiring the brain to break any negative messages a person may have allowed or embedded herself into her subconscious mind, for example by stating words aloud such as, "I'm in pain," "I feel sick," "I'm so fat," etc.

When a person spends all day or even just once or twice a day making these negative statements the person's subconscious eventually becomes confused causing imbalances, which can result in pain and emotional turmoil. This is when wires can become crossed, as previously discussed, and thus your intuition might direct you astray even to the point of making those statements come true. Conversely if you regularly make positive statements, such as, "I feel good," "I feel strong," and "I'm healthy," regardless of your actual physical state the subconscious mind receives those messages and realizes it must make those things happen.

First, the subconscious mind seeks balance and, second, it seeks the best for you thus tending toward success, therefore it is going to direct you to choices, which will allow you to achieve whatever positive messages you are embedding into your subconscious. In essence, by making positive statements, whether or not you actually believe them, you are directing your conscious mind to be in sync with the subconscious, which is the condition most conducive to success.

It is not a miracle cure nor does it happen overnight; the reasoning for why and how this works is quite logical but requires an understanding of the things discussed in this book. As stated, the subconscious mind communicates with you through your intuition. Your subconscious mind knows what is best for you; however, if your conscious mind and subconscious mind are not on the same page the messages sent through your intuition might be twisted.

An interesting question is, "where" is the subconscious mind located? Where the subconscious actually resides in the body is not readily available in the scientific literature, however there are some theories and even solid beliefs by certain types of practitioners. Some people believe it resides within the brain, others believe it is within the heart. Could it be just a concept? Is it logical that the subconscious mind actually exist as part of your brain, and if so, where? Perhaps it is that part of the brain of higher intelligence, e.g. the part where rational thought exists.

The prefrontal cortex is responsible for instinctively reacting to things in daily life. It is the part of your brain that directs you toward your goals. Based on what we know about the subconscious mind it appears that it might reside within the PFC, or actually *is* the PFC; however, this could be simply the location of the conscious mind, as those who experience "higher" consciousness actually experience increased activation of the prefrontal cortex.

Seventeenth century philosopher, scientist, and mathematician René Descartes referred to the pineal gland as, "the principal seat of the soul." This sounds a lot like a description for the subconscious. Being located near the center of the brain he could have been correct but who is to say? The subconscious mind is an elusive topic in the scientific world. Other doctors and philosophers, including those from India, believe the subconscious resides within the heart. What if the subconscious mind is simply the soul itself, which flows throughout the entire body? These are interesting questions, to say the least.

Wherever the location, it is clear the conscious and subconscious minds are connected and thus there must be some sort of communication between the two beyond intuition, something potentially quantifiable and visible through brain scans. If a connection itself is not visible through scans perhaps communication can at least be seen as it occurs.

In fact, one study conducted on individuals performing affirmations shows that the prefrontal cortex of the brain is activated when doing a certain type of affirmation. In particular, the individuals wrote down affirmations in response to a questionnaire. The study showed that the prefrontal cortex is activated when the written affirmations relate to a concrete future event such as achieving a particular positive goal or reward. If an individual imagines, and documents, a goal he or she would like to accomplish in the future the prefrontal cortex shows a great amount of activity. The same is not true when the individual imagines common daily tasks in the future. Speaking, versus writing, the affirmations was not part of this particular study; however, it would seem that speaking positive affirmations of a future goal, such as being totally healthy, or even obtaining something like a better income or nicer car, would make that goal more likely to come to fruition than if one did not speak, let alone think, these things.

The good news from the study, however, absolutely shows that written affirmations of a positive future goal triggers the PFC and therefore your higher intellect and rational brain is going to work toward making that goal happen. It was like the gentleman I previously mentioned who writes seven daily affirmations in the first, second, and third person. Indeed it is a habit we should all consider creating. Studies have shown that people who write down their goals or dreams earn nine times as much as those who do not write them down.

Ultimately, it is imperative that we not feed our subconscious mind negative emotions or feelings, as the subconscious will either be confused and send the wrong messages back through intuition hampering our goals of health or even wealth, or the subconscious will do what it must to make the negative idea come to realization. Speaking in positive tones and outwardly proclaiming our goals is effective, but written goals have actually proven to be highly successful.

Many who subscribe to new age beliefs seem to promote this concept as, "you can manifest whatever you wish" or " you create your own reality". In my view it is not so simple as thinking something and POOF it appears. When you provide your subconscious mind with a goal it will create a blueprint for achieving that goal and will trigger the rational part of the brain to spring into action. This does not mean it is performing magic and will manifest your goal one day even though all you do is watch TV all day long. Instead, it will direct you to take action toward your goal. It will use the blueprint to direct your brain toward achieving the goal, which will eventually, one day, be realized. I do believe you could sabotage your goal by confusing your subconscious mind with negative ideas or worries and fears. If you imagine you will fail you could affect the blueprint so that you do fail.

What does it all mean? It means there are things you could do to effect positive outcomes in your life. First, practice deep breathing; second, always try to direct your thoughts to the positive and catch yourself if you begin to imagine a negative outcome or have fear or worry; third, write down your goals as affirmations and do this daily or as often as possible. It is all about habit creation and once you have developed the habit of deep breathing and writing, or even speaking, affirmations it will eventually stick.

While it is quite simple it is not easy. Habits are not easily formed or broken; however, with some discipline it can be done. It has been found that performing a habit daily for twenty-one days it can be formed. Consider this when aiming to effect positive change in your life.

Remember it all stems from that invisible force, energy. Like the wireless router in our homes sending invisible signals to our computer, TV, and other devices, the messages sent to the prefrontal cortex are accomplished through electrical signals along the nerves. Energy is required for this electricity to do its job. It is *all* about energy. The healthier your brain, especially that prefrontal cortex and the more it is in sync with the subconscious mind, the more effective the messages will be.

Chapter 24

Schumann Resonance
The Earth's Heartbeat

I had initially intended to include a chapter on this topic in this book and then abandoned the idea as being too complicated or not relevant enough to the topic at hand to warrant great detail. This is until I came across a YouTube video called *The Schumann Resonance Bursts and Affects on Human Consciousness* from the channel "Destroying the Illusion". The owner of this channel creates short videos in which he discusses many esoteric topics tying scientific concepts to the discussion in such a way that makes the information highly credible and understandable.

Call it coincidence, the law of attraction, or perhaps the "Universe" in tune with my subconscious mind but the suggestion popped up on my recommended videos without any prompting. The video itself had been released several months before I saw it. I was obviously drawn to watch, as I had been debating on whether to write about it or not. The information in the video itself was on a topic I had not considered as relating to the Schumann Resonance, however it was relevant to this book even more than I had originally believed.

For many years now I have heard people, including myself, making the statement that time seems to be going faster. I believed it was simply that as we get older and become busier time just seems to go faster; I assumed it was that the older you get the faster time goes by due to our activity, as though we led a simpler life as children and literally had "lazy days". Of course I remember playing. A lot. Yet time did not fly as it does today. If I would ask for permission to play with my cousins, my mom would say, "Okay, for an hour", and sometimes only half an hour. A half-hour was plenty of time to get some good fun, an hour an eternity.

These days I am stunned at how quickly the weeks go by. Nearly every Friday morning as I am about to get out of bed I think, "It was just Monday. It's Monday and it's Friday, it's Monday and it's Friday." It feels like there are two days a week rather than seven.

Every year for Christmas I make my home festive by decorating with garland and Christmas lights around the wooden vigas that are atop the half walls surrounding my sunken living room. Occasionally I will add lights around the windows if I have time. There's that word again. The past few years as I have done this, as I am standing on the ladder hooking the lights around the posts I think, "Wow, I feel like I just did this. Before you know it I'll be taking these decorations down and it will be spring again." I have been saying for a few years now, maybe since around 2012 or 2013 that, once June rolls around, Christmas is here before you know it. I know that sounds nuts but that is how it is. It really seems like insanity.

I remember a few years ago being in the women's locker room of the gym at my work and making a comment to a woman about this very thing. What she said struck me. She said, "I agree and I always thought it was just adults but my children are saying the same thing." At that moment I knew something was going on. That evening I began researching the phenomenon and discovered the concept of the Schumann resonance. This was several years ago and I no longer have access to that video, which I might be able to find with some digging, however I recall that the engineer discussing the concept stated it was not until around 1984 when it appeared the Schumann resonance began increasing and, since then, it seems the passage of time has also begun to increase. According to this engineer it was attributed at the time to the increase in electronic devices and wireless technology.

In the 1950s a physicist named Otto Schumann discovered that the earth had its own frequency, which was eventually termed the Schumann resonance. This frequency is considered to be the earth's heartbeat. The number measured by Schumann was 7.83 Hz, which is believed to have been the lowest frequency. Of course, if that was the first time it was ever measured, who knows? Perhaps it has gradually increased over the centuries, it just wasn't measured.

Jordan Sather, the journalist behind the Destroying the Illusion channel, claims that in 2014 scientists noticed the resonance spiking to frequencies between 11 and 14 Hz. It seems to me this is around the time I really began feeling that time was speeding up. Around Easter of 2017 it was observed to have spiked to an all time high of around 90 Hz [Author's comment: since then it has spiked even higher]. I believe it is important to dig deeper into this subject and the concepts discussed in Mr. Sather's video because his observations and research make perfect sense as to where we are today in terms of our connection with others and ourselves as well as our consciousness, which is a main theme in this book.

In Sather's video he discusses the concept of consciousness and how it is related to the Schumann resonance. According to his research human consciousness is tied to the Schumann resonance and our consciousness increases as the frequency of the Schumann resonance increases. He claims the sun is going through an evolution, which is causing spikes in the Schumann resonance and these spikes are responsible for causing great anxiety within people, particularly those who are unaware of this connection and unable to easily adapt. He further states this anxiety and mental turmoil is actually a result of an increase in consciousness, which is necessary for our own evolution.

An increase in consciousness simply means we are thinking much more, rationalizing, inventing, solving, etc. The higher the consciousness the more understanding we have but if we do not allow ourselves down time through just being silent, or using meditation and deep breathing techniques, we could be at risk of having great turmoil in our minds that could become full-on anxiety. Moreover unhealthy habits such as not getting exercise, having poor sleep, poor nutrition, and exposure to toxins contributes to the anxiety. Not leading a balanced lifestyle can contribute to that anxiety, as we might be allowing our minds to "think" too much. Further, a lack of understanding of what is happening could also be detrimental and many people might squander an incredible opportunity.

Sather does not blame anything manmade for the sun's evolution and indeed it might not be due to anything manmade but there are others who believe the Schumann resonance has increased because of the pervasiveness of electromagnetic pollution in existence today, which did not exist in the past. This would actually stand to reason and is tied in with the earlier discussion on the importance of a good night's sleep, which can be disrupted by the existence of unnatural electromagnetic energies and devices.

As a side note, which should be mentioned here, scientists have determined that the measurement of the Schumann resonance has become very difficult to achieve due to the pervasiveness of electromagnetic pollution. Physicist Dr. Wolfgang Ludwig has stated, "Measuring Schumann resonance in or around a city has become absolutely impossible. Electromagnetic pollution from cell phones has forced us to make our measurements at sea." Thus even if the electromagnetic pollution itself has not affected the earth's heartbeat, it certainly has affected its measurement. It would be difficult to imagine it would not be affecting the frequency itself.

The reason physiological disruptions are tied to the Schumann resonance and impacted by electromagnetic pollution is because the body seeks to synchronize itself with the natural energy with which it is surrounded. The body has adapted to the energy coming from the earth, which is tied to the sun, the moon, and other celestial bodies. Society's level of consciousness as related between the earth's heartbeat and the sun's energy can purportedly be quantified in terms of mental anxiety or the acceptance of and adaptability to increased consciousness. As previously stated, studies have shown that people can predict geomagnetic disturbances based on their state of mind and physical symptoms.

Of course the idea that the spike in Schumann resonance is tied to consciousness is not fully vetted or proven. Some scientists who monitor the resonance and who study its connection with living beings claim the spikes have occurred before and it is not unusual. So, while there is debate as to whether or not the spikes in the Schumann resonance has any bearing on consciousness it seems safe to say we are indeed connected with one another, possibly even on a level of consciousness.

One study in particular confirmed our connection as living beings to the earth and her heartbeat. Through heart rate variability testing it was found that people living in various parts of the state of California showed a distinct change that coincided with solar activity as well as changes in the Schumann resonance. Although the same was not evident with regards to the indexes used to measure geomagnetic storms, it was clear that the body is constantly seeking to synchronize itself with nature and thus the sun. By the end of the study not only had the heart rates of the participants synchronized themselves with the earth's frequency but they also synchronized themselves with each other. The results were closely corroborated with other studies conducted on larger groups of people over larger geographical areas.

The implications of this are huge if you really think about it. Because this is all tied to consciousness and the idea that our hearts are connected with each other seeking to beat as one with the earth, imagine the effects we each have on one another. It is a beautiful and scary thing. Now, more than ever, we must understand how we individually have the power to change the world first by recognizing this connection and, second, by ensuring we do not allow negativity into our space.

A certain type of meditation or, "mind control" as it were, recently came to my attention. Again, it is interesting how I have attracted certain individuals into my space and even certain concepts just by working on this book and by having these topics on my mind. The "Silva Mind Control Method" is one such concept. This along with other various methods have been developed over the years, which allow a practitioner to quiet his own mind and control the frequency of the brain such that he is able to achieve, for example, an alpha state while being fully awake. Doing so shuts out all distractions so he is able to absorb information more readily and even essentially "read" the mind of another person. It is obviously a skill that should be used with caution but imagine the ability to connect with others on a conscious level simply by changing the energetic frequency of our brain.

In fact, getting into an alpha state is highly beneficial to activate the right brain, which is the creative side. By activating both sides of the brain, especially the right brain it is said we are able to activate the higher consciousness written of in this book. Doing this can allow for great possibilities, not only manifesting our goals and dreams, but also of achieving great healing; physically, emotionally, spiritually, and beyond. Moreover, tapping into the right brain elicits unimaginable creativity.

When I was still in high school I had a passion for writing both poetry and songs. The way I would find inspiration was by listening to calming, relaxing music. It was mostly during these moments I was able to create inspired poetry. After I started college, I would sit in my dorm room with music by Enya playing on my stereo and write poetry. I did this a few times until school became too busy and I was inundated with left-brain education. It was at this time when I felt I lost my creativity.

Little did I realize at the time that the simple act of listening to the gentle sound of music such as that by Enya would allow me to activate my right brain by lulling me into the alpha state. I assumed I was becoming more left-brain dominant due to my education and thus would have to bury any ideas of a creative future in the library of past knowledge of my mind. It was not until the drafting of this book that I realized I could tap into the right-brain creative side simply by listening to classical or new age music thereby entering an alpha state.

Many things can occur in an alpha state, only one of which is the tapping into of creativity. One of the most remarkable things that can happen in an alpha state is tapping into the mind of another. It is in this state that we are able to transmit healing energy, forgiveness, love, or even negativity to others in our midst.

The next time you are in a room, at an event, or in some situation with others in which one or more people are out of sorts or even confrontational remember this and know you have the power to change the situation for the better. The same is true with regards to conflict in our personal lives and within our homes. I believe if we practice tapping into our intuition, as previously discussed, and if we work on feeling our own energy, controlling our thoughts, deep breathing, and practicing mediation we would be able to affect the consciousness of others to the point of creating ultimate bliss through forgiveness, gratitude, and love.

It is difficult to step outside oneself and one's own circumstances and feelings to see others for who they are and what they might be going through. When someone hurts us with words or actions we can only see our own pain. It is easy to take things personally, especially when there is an attack on our personality or character and, even though we know deep down what the person is saying is not true, we cannot help but become angry and defensive. Instead of looking at the person and considering what might be causing the person's issues we engage in anger and conflict without any real rationale. One of the things I heard a long time ago, and a mantra I say to myself and to others in situations like this, is, "what other people think of you is none of your business." Indeed it only becomes our business if we allow it. I try to tell people not to allow other people's issues to become their own.

The other saying I believe, and to which I cling, especially when there is conflict, is, "we do not see things as they are, we see them as we are." I believe this wholeheartedly and the way I like to put it is that we see others as a reflection of ourselves. This means that when we have issues with other people it is because it is what we ourselves feel. For example, if you are offended by certain words said by someone in your life, it is likely because you would use those same words in an offensive way. When we become defensive by the words of others there is usually a shred of truth to it.

It is safe to say most people just want to be happy and do not wish to hurt others. Therefore when someone accuses you of hurting him or her it is important to approach the situation from a place of compassion and recognize it is not personal but the person projecting his or her own issues upon you and you have a choice of engaging and allowing it to bother you or to use your power of connection with the collective consciousness to change the situation by projecting love and understanding.

Perhaps the person who has said hurtful words is struggling with his or her own increase of consciousness and does not understand what is happening. Perhaps memories of past hurt are resurfacing creating turmoil in their minds. Whatever the situation it is more important than ever to use our own consciousness to prevent further conflict and pain.

It is profound in our own personal lives. When we become offended by what others say or do, especially when it is apparently a benign, petty issue, it is important to self-reflect and work toward healing whatever may be broken inside that caused us to see a situation in such a way. You may be thinking, "Oh yes, I know this and I've been there. I have been able to see this in my own situation so I'm good", but we must never rest on our laurels. There is an ebb and flow to this, as sometimes we are strong and secure within ourselves and other times we take things much too personally. Life is a constant quest for growth and as our circumstances change, such as experiencing loss or hardship, we can become wounded, which can in turn cause us to be too hard on ourselves. Before we can forgive others we must forgive ourselves and before we can love others we must love ourselves.

Maybe, just maybe, if enough people are able to self-reflect, beginning the process of forgiveness of themselves, and ultimately develop a strong self-love it might affect the collective consciousness thus those who need to love themselves the most will begin to feel a stirring in their soul to dig deeper and try to understand this. The ultimate goal, of course, is that we all become capable of forgiveness and eventually loving ourselves fully. Maybe when this happens, as our hearts beat as one, we can see conflict for what it is, an underlying pain from the disappointment in ourselves and as one person loves himself and thus is able to show unconditional love for another person, our hearts and souls might begin to heal as one and harsh words are understood to be a calling for even more self-love and reflection. The earth herself is here to show us and to help us connect to one another. We must not squander this gift.

Chapter 25

What Does It All Mean
and How Do I Put It Into Practice?

At this point either your mind is reeling from everything you have read in this book and you feel you might need to go through it again to fully understand the information; you are feeling justified in believing in "the law of attraction" and how energy is such a big part of our lives; or you're wondering how to put it into practice to help you improve your life. Although I have already given some ideas on how to do this there are a few things that should be summarized for some clear direction.

The beauty of energy is that it will never go away. Remember the law of conservation of energy: "Energy is neither created nor destroyed". It is always there and therefore you will always be able to tap into it for inspiration or to change your life completely. Unfortunately, when it comes to the energy we hold, if our bodies are not functioning optimally we are squandering opportunities to use that energy to our benefit.

In terms of holistic functioning, we should be able to use energy in every aspect of our lives so as to be completely well rounded, perfectly content and ultimately happy. Even adversity is viewed as a positive opportunity and challenges are welcomed. This might sound like a pipe dream but it is actually possible. The difficulty arises in being out of tune or out of sync with your inner self.

Fortunately there are tools available for you to get in touch with your spiritual/energetic side, as previously discussed. Where does one begin? This is not a small feat, as one might imagine, but some people are further along in the process than others whether or not they realize it.

One very good way to begin getting in touch with your energetic "wiser" self is by meditating. Many people believe meditation is a strange new age practice done by people who worship strange gods. In fact meditation is simply a way to connect with oneself by focusing on one thing such as breathing while shutting out all the craziness of life. Indeed connecting with yourself means connecting with your spirit and thus there is spirituality involved. We are, after all, spiritual beings. We must not get lost in semantics; instead we must understand how important it is for one to connect with oneself and, thus one's spirit.

It is not necessary to deep breathe for half an hour or even for a whole five minutes. In fact, just taking a few deep breaths can get a person into a meditative state deep enough to begin connecting with oneself. The reason this is so important and why it works has its roots in science and physiology.

As previously discussed, when we breathe deeply, especially when the spine is healthy, the breath makes its way to the prefrontal cortex, which is where our higher selves and our highest minds exist. This oxygenation stimulates this area and invigorates the cells. This is the part of the brain responsible for so many things, including rational thought. Thinking rationally will help you become more grounded to realize that whatever problem you face is really not very big and there are solutions. Like the saying (and book) states, "Don't sweat the small stuff...and it's all small stuff."

Unless you are ready to go all in, once this becomes a routine practice it is a good idea to graduate into other areas of energetic presence. This could be as simple as incorporating the tapping involved in EFT, as discussed earlier, or taking it a step even further and using the arm length test to start asking your own body questions. If you choose to practice the arm length test it is highly recommended to become familiar with Innerwise and understand the method to begin asking your subconscious questions so as to ensure you are receiving the most accurate answers. Again, there are free videos on the Innerwise web page at Innerwise.com to assist you.

If you find this to be very difficult chances are you must get back to basics and learn simple meditation techniques. You could seek guided meditation programs if necessary. If you find you are simply "stuck" and nothing seems to help you connect with yourself you could seek out practitioners who specialize in energy healing, such as those involved in NSA and SRI, acupuncturists, homeopaths, or other doctors who use energetic concepts such as Dr. Storkan.

Be advised and forewarned, however, that not all practitioners have their techniques rooted in science. While certain forms of prayer can help a person connect with the spiritual, other forms such as those particularly rooted in New Age or pagan rituals should be approached with caution. Before delving into anything like séances or even using a "game" such as a Ouija board one should be fully aware of the implications and what others have experienced including the inadvertent opening of portals to other dimensions. As referenced before there is such a thing as negative energies, which can be invited into a person's space and will wreak havoc without you realizing it. For this reason it is recommended to conduct some research before participating in any type of energetic practices, including certain energy healing modalities.

When I write about these things even I feel like I sound a bit silly and superstitious, however, as someone with a background in science who has done a great deal of research into the scientific aspect of energy with regards to the human body I recognize the danger in a lack of understanding. It is my belief this is where the teaching within the bible, which forbids dabbling in divination, originates. I do not believe it is so much that energetic practices are always "bad" but because the danger is in lacking understanding. Better to be safe than sorry.

In fact, during the workshop with Dr. Albrecht he described a situation in which some students from a European country invited him and his wife to their home and said they had a "surprise" for him. He had no idea what the surprise would be but, to his dismay, the couple had arranged a séance. Needless to say he reminded the couple that his practice only involves the healing of oneself either alone or through a practitioner and conducting such rituals was a very dangerous practice and should be approached with extreme caution.

I would like to convey the message that energy is arguably the most profound and strongest aspect to our lives and existence but probably the least understood or utilized. I wholeheartedly believe we all would benefit from both understanding and using it daily to assist us in not only becoming healthier but also in improving every aspect of our lives including relationships, finances, family, etc.

Remember our bodies are always tending toward healing and our subconscious minds want the most success for us. As previously stated, many people believe we are connected with our "Source", the "Universe" or God, however we choose to view it, directly through our subconscious minds. God wants the best for us and all humanity thus as long as we pay attention to it and not allow our logical mind to get in the way then we will have the ability to have joy, love, forgiveness, and understanding, even in the face of adversity. Adversity itself would be viewed as a gift. For without knowing sadness we cannot know joy. Without feeling pain, we might not appreciate feeling good.

When thinking in terms of our subconscious mind and our "Source" we should consider the gift of free will. It is when our free will is in sync with our subconscious, or the will of God, that our bodies are most in balance and when we are able to speak the language of our subconscious mind through intuition. One of the most powerful tools in our healing toolbox is to follow our intuition most effectively. I cannot stress this enough. As I previously stated, we are living in confusing times especially when it comes to our health and the recommendations we have received from the mainstream.

Science is not definitive. It is constantly evolving and new concepts are always being discovered. Medieval practices such as blood letting were once believed to be beneficial and at one time cigarette smoking was considered healthy. It is important to be open to new concepts, or even the renewal of older concepts, regarding health. However, it is also important to realize we are each unique in terms of health and our bodies. Even recommended daily allowances of certain nutrients might be very different for one person than they are for another or the accepted ranges of various measurements in the body, as provided by blood work, might not be applicable to all individuals. For these reasons having the ability to tap into your intuition to ask your own body exactly what it requires is extremely beneficial.

Above all, as energy relates to thoughts and belief systems we must be mindful of how powerful those belief systems become in our lives. Remember the discussion on how doctors might inadvertently shorten their patients' lives by giving them a deadline. The reason the person sometimes meets their fate along the timelines created by the doctor is due to faith, or belief. In my view, regardless of the course of treatment, our belief system has the power to heal us or even shorten our lives further.

Even if a person is undergoing the most ridiculous, toxic therapy, if he or she believes it will work then it is highly likely that it will work. Thus, in some instances it is best not to insert any conflicting information or negativity, which might shatter a person's belief system to the point of hindering her healing. Natural treatment modalities work the very best when a person believes in them. I believe wholeheartedly that a person's energy, triggered by her thought process, rooted in her belief system can counteract the harshest, most toxic of treatments if her faith is strong enough. Indeed it takes a very faithful individual to accomplish this but nonetheless it is possible.

Understanding our connection with others is also highly beneficial in terms of dealing, especially, with conflict. Recognizing this connection as one based on energy is critical. Moreover, realizing there are unseen forces affecting us for better or worse is also imperative in order to ensure success on all levels. If the messages sent from our brains through our nervous system via electrical signals can foster complete healing, imagine how those same signals could sabotage our health if we are not energetically balanced.

Finally, I felt it was critical to include the information in this book regarding the dangers of artificial frequencies and how they can affect the health of our bodies and mind. Even natural frequencies coming from thoughts, words and intentions can cause great damage if we are not careful.

Consider a popular experiment, which could be viewed as an offshoot of Dr. Emoto's work on water crystallization, where three separate jars of rice are observed over a period of time. Each jar of rice is either exposed to beautiful, positive words and intentions; horrible, negative words and intentions; or ignored altogether. Of course it always goes back to intention but if the experiment is done correctly there are three distinct results. As you might imagine, the rice, which is exposed to ugly words and intentions becomes rotten and putrefied while rice exposed to beautiful words and intentions retains its freshness for quite some time. What you might not imagine is that which is ignored altogether also exhibits negative results, namely mold.

The best illustration and discussion regarding this phenomenon I have found is on YouTube at the channel of Truthstream Media called "The Power of Your Intentions (The Rice Experiment)", which was uploaded on January 4, 2018. I strongly encourage a review of this video, as it illustrates much of what I have discussed regarding the power of thought and energy that is associated with it.

Although there have been no formal studies on the effects of sleeping next to a cell phone or in the same room as your wireless router just having a basic understanding of energy and frequency we are able to surmise that unnatural frequencies might disrupt our cells, whether or not they disrupt natural frequencies. Although there have been claims of informal school science experiments, what we know about the damage potentially caused by cell phone signals and actually caused by cell phone signals, which have been studied scientifically, and even the anecdotal cases of people who physically feel the effects of heavy wireless activity such as from smart meters it is safe to say we would be much better off limiting our exposure.

Considering that positive intentions can result in beautifully crystallized water and preserved rice it is clear we have the power to clear negative energies from our space and even protect ourselves from rogue signals simply by what we think. Remember this the next time you become upset with someone and, especially, with yourself. Self-hate is not just affecting your psyche it is also affecting your health. Make it a point to love as much as possible and to speak words of love as much as possible to others, and especially yourself. It will not only make you feel happier, it will make you feel healthier.

With all of this in mind it is not something to take lightly and anyone who utilizes energy healing would get much more out of it with a good understanding of energy both in the spiritual sense and also the scientific. Having this understanding could allow people to transcend many things, including petty arguments related to politics, materialism, and other differences of opinion.

A gap has been continually growing between people footed in the reality we "see" and those who understand there are other, invisible, forces affecting that "reality." It is time to bridge that gap with education in order for society to understand the divisiveness in the political and socio-economic system so we might rise above and take control of our wellbeing on a global scale. Without a basic understanding of energy society is doomed to live the same nightmare over and over again and continue to allow needless wars and suffering.

Afterword

It should be obvious that although I have included information on energy obtained from actual scientific studies some of what I have discussed comes from my own interpretation and my opinion. Many of these concepts are complicated in the sense that energy is everywhere thus connecting the dots between two points is not always straightforward. This is also why tapping into intuition is critical. As described in the quotes by famous individuals, such as Nikola Tesla and Steve Jobs, which I discussed in Chapter 16, this book was written from my heart, my rational mind, and, especially, from my intuition. The book has evolved from a very small project into something I now realize is more relevant to our time than I had initially thought.

During the drafting of this book many concepts presented themselves to me in unexpected ways. It was as though the Universe, or God, as it were, provided me the information I needed as it was needed.

As I write, I allow the words to come freely and only during editing have I "second guessed" myself. Because of the nature of intuition and the fact that the first choice is the best choice I have tried to be careful not to change information during editing, and only modify for better grammatical flow.

This book is, by no means, intended to be definitive on the subject, as there is much to understand and recognize with regards to energy and how it affects our lives. It is intended to offer a basic understanding of energy such that it will cause the reader to be more aware of his or her thoughts and intuition.

There are additional concepts I did not discuss, which are also highly beneficial to understand. These concepts include but are not limited to topics such as grounding, sound and music as it relates to healing and even destruction, light healing, symbology and the energetic meaning of such images as logos, numerology and the codes hidden within sequences of numbers, and the related topic of mathematics and concepts such as the string theory. I only touched on some of the healing modalities available, which utilize energy.

There are other topics concerning energy, which I feel are interesting to look into, if not critical to understand, namely nutrition and how our bodies convert the energy in the food we eat into essential energy for our bodies to use. As all things are energetic, everything, including both food and toxins, has its own energetic vibration and frequency, which can contribute to health or illness.

Additionally, a deeper insight into the collective consciousness and how we are all connected to one another might prove to be invaluable in these times of chaos. Understanding this energetic connection through a scientific lens can also support a deeper insight into the spiritual world. Summarizing the lectures, workshops, and other talks by exorcist priests such as Father Ripperger has the potential to provide great enlightenment in terms of personal issues that cannot be explained by science or medicine. Finally, there are other healing modalities of which to either embrace or be cautious. It is highly recommended that you browse the Appendix of this book and seek out additional information on the topics presented herein.

For these reasons I have made the decision to write a series of books on *The Secret of Energy*, which will delve much deeper into the topics discussed herein. My primary goal for this book has been to present to the average individual not only the prospect of energetic concepts but also their importance in our lives. Many people, including medical doctors who are actually trained in science, dismiss energetic healing as nonsensical or lacking all benefit when in fact it plays a critical role in our lives. Now, more than ever, we are faced with such things as electromagnetic pollution, brain tumors caused by cell phones, and smart meter sensitivity. These harmful and potentially deadly concepts are not going away anytime soon therefore it is more critical than ever to understand energy and how to protect ourselves from the damaging effects of unnatural frequencies.

My web site, http://secretofenergy.com will be a central location for not only this book in all formats – paperback, electronic, and audio, but also for future books. It will also be a hub for useful tools and items used in energetic healing, clearing negative energies, redirecting rogue signals, and locating certain practitioners of energetic healing.

What you have read should be a basic guide toward better health and wellbeing allowing an understanding of energetic healing concepts such that you might be much more open to seeking this type of assistance on your journey through life. Energy is not something we should ever fear but something we should embrace with understanding.

Above all we must be mindful of our thoughts, actions, and disposition remembering how they not only affect our own success, or lack thereof, but also those around us. Recognizing, and utilizing, the many energetic tools we have available can lead to immense possibilities. The most important and powerful tool in our arsenal is our intuition. If only we would pay attention to our intuition we could change the world.

If there is a single message I wish my readers would receive from this book it is the importance of intuition and how it is critical to the success of all that we pay attention to the messages our bodies give us on a daily basis. Suffering, as it presents itself through pain, sadness, and other physical and emotional symptoms is, in essence, your intuition speaking to you loud and clear that there is disharmony within your body and should be addressed, rather than masked. And, remember, protecting your subconscious is paramount.

If you are able to work toward tapping into your intuition you will have the world at your fingertips. You would be able to determine the foods that are right for your body and nutritional needs. As stated before, we are each unique and although there are certain nutrients commonly required by all there are others who are sensitive to certain foods, even nutritious foods. This is akin to food allergies. First, understanding how to tap into your own intuition you would have the ability to avoid the foods that might upset you. Second, if you are able to use energetic healing, such as through Innerwise or practitioners like Dr. Storkan, you could clear the energy pathways so you are no longer sensitive to those foods.

With regards to overall success, using your intuition to guide you toward the most effective decisions, you would find pervasive opportunity. Your subconscious mind would direct you along the most successful course and it would create a highly beneficial blueprint for your life. Never forget the importance of positive thoughts and faith. Never doom yourself to failure. Think in terms of abundance no matter your situation. Believe in what you can achieve and see it happen. Imagine the possibilities!

Be mindful of what or whom you allow into your space. Remember again, just because you cannot see energy does not mean it doesn't exist. Here's to healing our bodies, our hearts, our souls, and changing the world. Blessings on your energetic journey!

Appendix
References and Resources

1. Allen, Marshall, Pierce, Olga, Medical Errors Are No. 3 Cause Of U.S Deaths, Researchers Say (2016), https://www.npr.org/sections/health-shots/2016/05/03/476636183/death-certificates-undercount-toll-of-medical-errors.

2. Becker, Robert O., Selden, Gary. (1985) The Body Electric: Electromagnetism and the Foundation of Life. New York: Morrow.

3. Raible, F., Takekata, H., and Tessmar-Raible, K. (2017) An overview of monthly rhythms and clocks. *Frontiers in Neurology* 8:189.

4. Reinbert, A., Smolensky, M., Touitou, Y. (2016) The full moon as a synchronizer of circa-monthly biological rhythms: Chronobiologic perspectives based on multidisciplinary naturalistic research. *Chronobiology International* 33(5):465-79

5. Kronfeld-Schor, N. et. al, (2013) Chronobiology by moonlight. *Proceedings. Biological Sciences* 280(1765):20123088.

6. Brjigin, J., Zhang, L., and Calinescu, A. (2012) Circadian Regulation of Pineal Gland Rhythmicity. *Molecular and Cellular Endocrinology* 349(1):13-19.

7. McCraty, R. et al. (2017) Synchronization of Human Autonomic Nervous System Rhythms with Geomagnetic Activity in Human Subjects. *International Journal of Environmental Research and Public Health* 14(7):770.

8. Aldrich, T., and Easterly, C. (1987) Electromagnetic fields and public health. *Environmental Health Perspectives* 75:159-171.

9. Bogdan, L. et al. (2014) Influence of Electric, Magnetic, and Electromagnetic Fields on the Circadian System: Current Stage of Knowledge. *BioMed Research International* 2014:169459.

10. Korkmaz, A., et al. (2009) Melatonin: An Established Antioxidant Worthy of Use in Clinical Trials. *Molecular Medicine* 15(1-2):43 – 50.

11. Redlarski, G. et al. (2015) The Influence of Electromagnetic Pollution on Living Organisms: Historical Trends and

Forecasting Changes. *BioMed Research International* 2015:234098.

12. Carlberg, M., and Hardell, L., (2017) Evaluation of Mobile Phone and Cordless Phone Use and Glioma Risk Using the Bradford Hill Viewpoints from 1965 on Association or Causation. *BioMed Research International* 2017:9218486.

13. Hardell, L., Carlberg, M., and Hansson Mild K. (2013) Use of mobile phones and cordless phones is associated with increased risk for glioma and acoustic neuroma. *Pathophysiology* 20(2):85 – 110.

14. http://www.earthcalm.com/are-smart-meters-really-dangerous

15. Thomas, William. "Wireless, Chemtrails, and You" *GeoengineeringWatch.org.* Geoengineering Watch, 28 February 2013. Web. 4 December 2017.

16. Burrell, Lloyd. "5G Radiation Dangers – 11 Reasons To Be Concerned" *ElectricSense.com.* Electric Sense, 12 May 2017. Web. 4 December 2017.

17. Goldsworthy, Andrew. United States. Federal Communications Commission *The Birds, the Bees and Electromagnetic Pollution.* Web: fcc.gov 2009.

18. Leppik, LP, et al. (2015) Effects of electrical stimulation on rat limb regeneration, a new look at an old model. *Scientific Reports* 5:18353.

19. Patil, S. et al. (2016) The Role of Acupuncture in Pain Management. *Current Pain and Headache Reports* 20(4):22.

20. Dooly, Timothy R. (2002) Homeopathy, Beyond Flat Earth Medicine. San Diego: Timing Publications.

21. Waisse, S, and Bonamin, L.V. (2016) Explanatory models for homeopathy: from the vital force to the current paradigm. *Homeopathy* 105(3):280-285.

22. Fye, W.B. (1986) Nitroglycerin: a homeopathic remedy. *Circulation* 73(1):21-9.

23. Hicks, Jerry and Hicks, Esther. (2006) The Law of Attraction: The Basics of the Teachings of Abraham. India: Hay House.

24. Hill, Napoleon. (1937) Think and Grow Rich. Cleveland: Ralston.

25. Innerwise.com

26. Reiche, E.M. et al. (2004) Stress, depression, the immune system, and cancer. *The Lancet Oncology* 5(10):617-25.

27. Denaro, N. et al. (2014) Cancer and stress: what's matter? from epidemiology: the psychologist and oncologist point of view. *Journal of Cancer Therapeutics & Research* 3:6.

28. Seyfried, T.N., et al. (2014) Cancer as a metabolic disease: implications for novel therapeutics. *Carcinogenesis* 35(3):515-27.

29. Oakley, Colleen. "The Power of Female Intuition" *WebMD.com.* WebMD, Archives. Web. 4 December 2017.

30. Sirota, Marcia. "How You Know He's Cheating: The Difference Between Women's Intuition and Paranoia" *HuffingtonPost.ca.* Huffington Post, 29 August 2011. Web. 4 December 2017.

31. Jensen, A.M. et al. (2016) Estimating the accuracy of muscle response testing: two randomised-order blinded studies. *BMC Complimentary and Alternative Medicine* 16(1):492.

32. Http://tbmseminars.com

33. Thomas C, Bhatia S. (2014) Cancer: brain-regulated biphasic stress response induces cell growth or cell death to adapt to psychological stressors. *Advances in Mind-Body Medicine* 28(3):14-21.

34. Albrecht, Uwe. (2012) Yes/No: Using the Arm-Length Test for Instant Answers and Wellbeing. London: Hay House.

35. Meneguzzo, Paolo et al. (2014) Subliminal versus supraliminal stimuli activate neural responses in anterior cingulate cortex, fusiform gyrus and insula: a meta-analysis of fMRI studies. *BMC Psychology* 2(1):52.

36. Liu, K.K.L. et al. (2015) Plasticity of brain wave network interactions and evolution across physiologic states. *Front Neural Circuits* 9:62.

37. Ripperger, Chad. (2015) "Fr Chad Ripperger - Spiritual Warfare Conference - Weapons Against the Demonic" *YouTube.com.* Sensus Fidelium, 5 November 2015. Web. 6 November 2015.

38. Ripperger, Chad. (2016) "The 6th Generation: Generational Spirits: Lost Generation to the One Current¯ Fr Ripperger". *YouTube.com.* Sensus Fidelium, 18 April 2016. Web. 20 April 2016.

39. Bobrow, R.S. (2011) "Evidence for a communal consciousness". *Explore: The Journal of Science and Healing* 7(4):246-8.

40. Emoto, Masaru. (2005) The Hidden Messages in Water. Hillsboro: Beyond Words.

41. Emoto, Masaru. (2007) The Miracle of Water. Hillsboro: Beyond Words.

42. Beauvais, F. (2008) Memory of water and blinding. *Homeopathy* 97(1):41-2.

43. Epstein, Donald. (2000) Healing Myths Healing Magic. San Rafael: Amber-Allen.

44. Epstein, Donald. (1994) The 12 Stages of Healing. San Rafael: Amber-Allen.

45. Falk, E. et al. (2015) Self-affirmation alters the brain's response to health messages and subsequent behavior change. *Proceedings of The National Academy of Sciences* 112(7):1977–1982.

46. Elster, J, Philos. (2010) Self-poisoning of the mind. Philosophical Transactions of the Royal Society 365(1538):221–226.

47. Taber, J. et al. (2016) A Pilot Test of Self-Affirmations to Promote Smoking Cessation in a National Smoking Cessation Text Messaging Program. *JMIR Mhealth Uhealth* 4(2):e71.

48. Cascio, C. et al. (2016) Self-affirmation activates brain systems associated with self-related processing and reward and is reinforced by future orientation. *Social Cognitive and Affective Neuroscience* 11(4): 621–629.

49. Dykes, Aaron and Melissa. (2018) "The Power of Your Intentions (The Rice Experiment)" *YouTube.com.* Truthstream Media, 4 January 2018. Web. 5 January 2018

About The Author

Marcie Martinez grew up the youngest of six much older siblings in the small community of Chimayó in northern New Mexico and has a great affinity for rural towns. At a very young age Marcie developed a passion for reading, writing, and helping others. She received a bachelor's degree in Chemical Engineering from New Mexico State University and a master's degree in Materials Science from Colorado School of Mines. Marcie worked for several years at the Los Alamos National Laboratory in various engineering positions, including nuclear materials science, plastics engineering, weapon test engineering, and quality engineering.

Being compelled to stand up for what she feels is right she has inadvertently been involved in various activist-type activities, which prompted her to declare herself an "accidental activist". Marcie hopes to share these experiences through her writing some day. Primarily she aims to share her research into natural healing to help others find optimal health and also to tap into their own passions using their intuition toward the goal of ultimate wellbeing.

In addition to her endeavors of researching and sharing information about natural health Marcie plans on releasing additional books on health as well as her own memoirs. To read more of Marcie's work, follow her at Medium.com @marciemtz.

Marcie also helps share her husband's beautiful metal art, which can be found at http://stmichaelmetalarts.com and http://facebook.com/stmikemetal. She considers herself his biggest fan.

Other contact information:

http://secretofenergy.com
http://naturespresence.net
LinkedIn: linkedin.com/in/MarcieMartinez
E-mail: Marcie@naturespresence.net